Ready-to-Build Telephone Enhancements

Delton T. Horn

TAB Books
Division of McGraw-Hill, Inc.
Blue Ridge Summit, PA 17294-0850

Notices

Touch-Tone AT&T and Bell Telephone

FIRST EDITION
FIRST PRINTING

Library of Congress Cataloging-in-Publication Data

Horn, Delton T.
Ready-to-build telephone enhancements / by Delton T. Horn.
p. cm.
Includes index.
ISBN 0-8306-4359-1 (paper)
1. Telephone—Equipment and supplies—Design and construction--Amateurs' manuals. 2. Electronics—Amateurs' manuals. I. Title.
TK9951.H67 1993
621.386—dc20 93-12883
CIP

For information about other McGraw-Hill materials, call 1-800-2-MCGRAW in the U.S. In other countries call your nearest McGraw-Hill office.

Acquisitions editor: Roland S. Phelps
Editorial team: Laura J. Bader, Editor
Susan Wahlman, Managing Editor
Joanne Slike, Executive Editor
Joann Woy, Indexer
Production: Katherine G. Brown, Director
Design team: Jaclyn J. Boone, Designer
Brian Allison, Associate Designer
Cover photograph: Brent Blair, Harrisburg, Pa.
Cover design: Joyce Thompson, Reston, Va.

EL2
4362

Contents

Introduction

We are now in an age of technology. We take so many modern inventions for granted today, even though they would have been considered absolutely miraculous not so very long ago. Some modern conveniences have become necessities, even though generations of human beings did without them before they were invented.

An excellent example of this type of invention is the telephone. A mere century ago few had even heard of such a thing. Today very few homes don't have at least one. And the telephone system is being put to use in new ways that were never even imagined by the original inventors. Today we have car phones, cellular phones, computer modems, answering machines, fax machines, and many other wonders of modern technology.

Early telephones were purely electromechanical devices, with few electronic components. Today most telephones and related equipment are fully electronic, making them a natural area for exciting projects for the electronics experimenter.

This book features more than two dozen add-on projects for your home or office telephone system. Most of these projects are quite simple and inexpensive. You will undoubtedly notice that these projects do not attempt to duplicate the functions of commercially available telephone-related equipment. There is little point in reinventing the wheel. A home-brew fax machine would be an overwhelming project for all but the most avid electronics hobbyist. That is, if you could find a source for the necessary specialized parts in single quantities and at reasonable prices. In the economic realities of the modern electronics industry, a large-scale manufacturer buying components in mass quantities and using mass-production assembly techniques can often turn out a completed product that costs significantly less than a home-brew project of comparable capabilities. At most you might save a few pennies. The commercial telephone-support field is quite competitive today. You can buy an answering machine for as little as $30 or $40, so there is scarcely any point in trying to build your own.

In choosing the projects for this volume, I looked for special functions that are either unavailable in commercial equipment or are available only in expensive equipment that might be overkill for the average user's personal needs. In other words, I tried to pick telephone-related projects that make good economic sense.

The first three chapters of this book provide important background information on telephones and the telephone system. Chapter 1 gives a quick overview of the history of the telephone. Chapter 2 discusses the modern telephone system and the types of signals that are used on the phone lines. Chapter 3 takes a look inside the instrument itself.

The projects begin in chapter 4, which features half a dozen remote ringer devices and related projects. Chapter 5 tells you how to add hold button functions to your home phone. The projects in chapter 6 are related to dialing functions. Both pulse and dual-tone multifrequency (DTMF) (or Touch-Tone®) dialing are covered.

In chapter 7 I look at several circuits that can utilize your phone for various remote control purposes. Telephone amplifiers are covered in chapter 8. As your phone system expands, it can become difficult to keep track of everything. The projects in chapter 9 will help you keep your phone system under control and secure.

Finally, sooner or later electronic circuits need to be serviced. Chapter 10 describes some simple but useful special procedures for testing phones, phone lines, and related equipment. Projects for building specialized test equipment for such purposes are also featured in this chapter.

With the projects presented in this book, you'll get more out of your telephone than ever. It will soon seem even more indispensible than before. That's the way things work with technological innovations. Something you never heard of yesterday becomes something you can't live without tomorrow.

❖ 1 The history of the telephone

These days it is difficult to even imagine life without the telephone. This device has been put to so many uses, and people have grown to depend on it. It is hard to believe that it's only been around for a little over a century. The official date for the invention of the telephone is usually given as 1876. Of course, as with most other technological inventions, development and research on the telephone began somewhat before this.

Interestingly, the telephone was really simultaneously invented by two independent researchers—Bell and Grey. Alexander Graham Bell is generally credited with the invention of the telephone because he was lucky enough to beat Grey to the patent office. Bell got his patent on March 7, 1876. If there had been just a few minor differences in history, we might have been speaking of "Ma Grey" instead of "Ma Bell." Somehow that doesn't have quite the same ring to me. (That pun was unintentional, but now that I am aware of it, it does seem appropriate.)

Bell's invention

Like many important inventions throughout history, the telephone began as sort of a mistake. It was a direct offshoot of experiments attempting to develop a device that could send multiple Morse code signals over a single wire. Today this concept is known as signal multiplexing.

The story goes that Bell was experimenting with a multisignal telegraph transmitter, while his assistant, Thomas A. Watson, monitored the receiver in another room. Bell spilled some acid on his trousers, and instinctively called for his assistant. "Mr. Watson, come here, I want you." In the heat of the moment, he forgot that Watson was not within earshot. To his surprise, Watson very quickly appeared—he had heard Bell's summons through the receiver.

Just how Grey stumbled on the principle of the telephone is apparently lost to history, but the basic principles of his invention closely paralleled Bell's. There is no need to go into the details of the minor differences between Grey's and Bell's telephones. Clearly the telephone was an invention whose time had come.

Bell's early accidental telephone didn't actually work very well, but it did work, and that was remarkable enough. At this stage, Bell's telephone wasn't very practical for widespread commercial use. For one thing, the system used corrosive acid—which is how Bell came to spill it on his trousers in the first place. In widespread use, such spills would inevitably be common occurrences, but the results would rarely, if ever, be as lucky as Bell's experience. Bell was lucky in more ways than one. He could have lost or seriously damaged his leg by spilling acid on it. Obviously an acid-based device like the early telephone wasn't likely to become a common household appliance, no matter how useful it was.

But some of the potential applications of this invention that could send voice communications over wires were obvious, and intensive research into the telephone concept soon followed. By 1878, the first commercial telephone exchange was opened in New Haven, Connecticut.

Ma Bell

The telephone filled a very definite public need, opening up long-distance communication without the need for cumbersome codes. Communicating by telephone was almost as easy as talking face to face. The telephone very quickly became big business. Poor Grey lost out on a lot more than fame.

One of the twentieth century's giant corporations was created specifically for this important new communications industry. The American Telephone and Telegraph Company (AT&T) was incorporated in March 1885 to oversee the rapidly growing telephone network spreading across the United States. Right from the start, AT&T had a rare opportunity to function as a legal, regulated monopoly. After all, a competitor would have to lay in their own complete set of telephone lines. It was assumed that competing companies would not (and should not) share their lines, so a customer who used a competing telephone service couldn't call someone on the AT&T lines, and vice versa. Later technological and legal developments eventually got around these problems, but the monopoly held solid for the better part of a century. The government began to express serious concern about this monopoly situation as early as the 1940s, but it wasn't until the mid-1970s that the monopoly was successfully broken up once and for all. The national telephone system was reorganized into a number of separate operating companies. Each still has a local monopoly, however. Which company you deal with depends entirely on where you happen to live. For example, in my area, the choice is to use U. S. West or do without telephone service altogether. This localized monopoly is simply a by-product of the nature of the technology. Company A is not going to let company B use their lines, any more than one store will give another store some of its stock. But it would be highly impractical to have two or three (or more) sets of telephone lines to every home and office in the area. Then there are the inevitable billing and connection problems when a customer of company A wants to call a customer of company B, when they are on different lines. To a large extent, some sort of monopoly is inevitable in the telephone business.

Right from the start, long before the breakup, there were hundreds of small local telephone companies not owned by AT&T. Generally AT&T was perfectly willing to let these independent companies install their own telephone lines and provide service on the local level. For the most part, these were small, mom-and-pop operations, mostly in rural communities with relatively low population density, where it wasn't really worthwhile for AT&T to install telephone lines themselves. But these little independent telephone companies still had to deal with Ma Bell to provide long-distance telephone service to their customers. Whenever anyone called outside the local community, AT&T got a hefty piece of the pie.

Whenever one of these local, small-scale telephone companies went broke, a far from infrequent occurrence, AT&T would step in and pick up the pieces, taking over the existing local telephone lines at minimum expense. Because the profitability of most of these local telephone services was rather low, many agreed to sell out to AT&T as soon as an offer was made. In this way, AT&T soon owned virtually all of the local telephone lines throughout the United States.

AT&T's monopoly extended beyond just providing telephone service per se. They had a strong hold on most of the communications industry, both in services and equipment. AT&T's equipment was manufactured by Western Electric, a subsidiary of AT&T. Because of the immense size of the corporation, they could afford extensive research laboratories. They remained in the forefront of the technology simply because nobody else could afford to do the necessary research on the level of Bell Labs.

The first major attempt to limit the size of the AT&T monopoly came in 1956. This was known as the *Consent Decree*. It allowed Ma Bell to keep her telephone monopoly on the condition that she would limit her business dealings to telephones. She was not permitted to dabble directly in any other business areas.

Ma Bell's first serious competitor entered the picture in 1969 when the FCC gave Microwave Communications, Inc. (MCI) permission to transmit long-distance telephone calls by air from St. Louis and Chicago. Gradually MCI was able to increase their range of operations, but at that point, AT&T remained the undisputed king of telephone services. Before the 1980s, few people had even heard of MCI, much less used their services. Today, of course, MCI is one of the leading long-distance services in competition with AT&T.

The next blow to Ma Bell's monopoly came in 1974. The U.S. Justice Department brought an antitrust suit against AT&T, on the grounds that they were freezing out all competition.

Deregulation

On January 8, 1982, Ma Bell's monopoly was officially broken. Ma Bell's assets were refinanced and restructured into seven regional companies, known as the "Seven Sisters" (that is, Ma Bell's "daughters"). The Seven Sisters are Ameritech, Bell Atlantic, Bell South, NYNEX, Pacific Telesis, Southwestern Bell, and U. S. West. In addition to the Seven Sisters, there is also Canada Bell, which continues to operate pretty much the same as it did before the 1982 reorganization.

The Seven Sisters are no longer directly owned and operated by Ma Bell. They function as a "cooperative" chain of independent companies. Ma Bell does, however, maintain control of her major profit maker—the AT&T Communications division. This is her long-distance telephone service. But Ma Bell now has significant competition in the lucrative field of long-distance telephone service, primarily from MCI and Sprint, although other, smaller long-distance communications services are in operation today.

AT&T also still controls the AT&T Technologies group, which includes AT&T Consumer Products, AT&T Labs, AT&T Technology, Information Systems, International, and Network Systems.

The AT&T Consumer Products division makes all sorts of electronic products for consumers, ranging from telephones to computers. AT&T Labs carries on AT&T's research and development programs pretty much as before. AT&T Technology is another manufacturing division, but this time on the component level, producing computer chips, circuit boards, and the like.

Information Systems is the revised version of the old telephone service conducted by AT&T. Worldwide communications services are provided by the International division. Finally, Network Systems is a renamed version of the old Western Electric manufacturing arm of AT&T.

The breakup of Ma Bell has had several significant effects on the ordinary consumer. For one thing, we now get to choose our long-distance service. The competition between the long-distance communications companies is fierce, forcing them to keep rates down. If company A charged significantly more for services than company B, most consumers would change over to company B, and company A would probably go out of business. I've heard a number of people blame higher telephone bills on the deregulation, but this is preposterous. A company with no competition can charge whatever the market will bear. A company with competition must consider the rates offered by its competitors and convince consumers they are the better bargain. Then why have rates for telephone services increased significantly since 1982? Well, what hasn't gone up in price in recent years? To a large extent, it is simply an effect of the overall troubled economy. In addition, the various long-distance services are in the process of converting to newer, improved technologies, such as fiber optics. In the long run, fiber optics will give better service at a lower cost, but right now we are paying for the installation of these new telephone lines, and that is inevitably expensive. Undoubtedly the cost of telephone service would have risen far more without the braking effect of competition, which did not exist before deregulation.

Another important difference deregulation has made to the consumer is that we can now own whatever telephone equipment we want. Before deregulation, AT&T owned everything and you rented your telephone from them for a monthly charge. You can buy a decent basic telephone today for $30 to $50. Compare this to renting. Let's say, AT&T charged you a monthly rent of $1.85 for the same phone. This certainly isn't much to pay per month, but it really adds up over time—$22.20 per year. After two years you have more than paid for the telephone (and remember, AT&T didn't have to pay any distributor or retailer markup for their telephones, just the bare manufacturing cost—probably about $5 to $10 per unit).

Actually it was never illegal to own your own telephone, although phone company personal often claimed it was, and most consumers believed it. But AT&T was able to make it extremely difficult and more trouble than it was worth for most consumers.

You can also add any additional equipment you might like to your telephone lines—such as answering machines, fax machines, remote ringers, and so forth. The only major restriction is that you are responsible for any damage your equipment does to the telephone lines or overall service. If you hook up something to your telephone line that disrupts your neighbors' regular telephone service, you can be held legally responsible. This certainly seems fair and reasonable.

Exact regulations vary from area to area. Generally you are supposed to notify the local telephone company of any equipment you add to your line. Commercial telephone equipment has a label with two numbers on it. One is the FCC registration number, which means the circuitry in the device has been checked out and approved for use on telephone lines—its effects are known. The other is the ringer equivalence number, which defines the amount of load the device will put on the line. This value is given in decibels, referenced to an old-style electromechanical ringer used in nonelectronic telephones.

As long as add-on equipment doesn't cause any problems, there will rarely be any complaints from the telephone company, even if you don't report it. However, I cannot guarantee this for your particular case in your paticular area. Local regulations vary quite a bit, so I can't be specific in this book. If there is any doubt in your mind, check with your local telephone company.

The author and publisher can accept no legal responsibility for the use of these projects.

The ringer equivalence for each of these projects is quite low, so any one of them should not put an appreciable load on the telephone lines. However, if you add a great many extensions and add-on devices, the cumulative loading effects might become significant. Again, when in doubt, check with your local telephone company.

Check out each project thoroughly before connecting it to the telephone lines. Check and double-check the wiring and all circuit connections. In some cases, a short circuit or error in the wiring could have disasterous effects. This is highly unlikely, but it could conceivably happen to a careless experimenter or hobbyist.

Unfortunately I cannot provide actual ringer equivalence values for these projects. That would require expensive equipment I simply don't have access to. In addition, the actual ringer equivalence can be affected by any component substitutions you might make, or even normal component tolerances. Therefore, I can't guarantee that my projects would have exactly the same ringer equivalence as yours. The best I can do is design each project to consume as little current as possible. Of course, these home-brew projects do not have an FCC registration number like commercially manufactured telephone equipment.

Overall, if you do no damage, you shouldn't face any legal problems for connecting projects like the ones in this book to your telephone line. It is the individual reader's sole responsibility to obey all applicable laws and regulations in his area. The author is not a lawyer and cannot offer any legal advice.

To return to the effects of deregulation on the average consumer, it is the individual consumer's responsibility to maintain and repair the equipment they own. In the old days, when you were renting your telephone from Ma Bell, you could expect free service. That is no longer the case. Most telephone companies will not even send a repairman out to deal with a broken telephone. They are only responsible for their own lines. In most cases, the telephone company's responsibility for maintenance ends where the wires enter your house. From then on, you are on your own. They will send service personal to troubleshoot problems with the wiring inside your home (or office), but you will probably have to pay for this service. A defect in the wiring outside is not your responsibility, and you should not be charged for such repairs, except in unusual circumstances. (For example, if you cut down a tree in your backyard and as it falls it knocks down some telephone wires, you will probably be held responbile for the damage you caused.) Most local telephone companies now offer some form of maintenance insurance to cover indoor wiring for a few extra dollars per month. Because it is usually a fairly simple matter to maintain the indoor lines on your own, this is a questionable bargain. Most consumers never have a problem with their indoor lines, but those that do might face a hefty service charge. You have to make up your own mind whether the maintenance insurance is a worthwhile investment for you.

Of course, electronics experimenters and hobbyists have a definite advantage in such matters. Some of the projects presented in this book are test equipment for checking out your telephone lines. This equipment should cover 99.99 percent of all problems you are ever likely to encounter.

❖2 How modern telephones work

Before you can intelligently examine any projects for telephone systems, you first need some background knowledge. This is the only way to avoid foolish and potentially disastrous mistakes. It is particularly important to know what you are doing when experimenting with telephone equipment, because if you accidentally (or deliberately) do anything to damage the telephone company's lines or disrupt service to any of their other customers, you could end up facing some very heavy fines and service charges, and possibly other legal consequences. Even if you feel you know all about electronics and telephones, I strongly urge you to read this chapter anyway. If nothing else, a review of background material has never hurt anyone.

To begin with, I will look at the basic electromechanical rotary dial-type telephone, even though these devices are rapidly becoming obsolete. Modern telephone equipment is based on extensions of the basic principles of the rotary telephone, so it is a reasonable and practical place to start. Even if you never come anywhere near a rotary telephone, the time it takes to learn about how it operates will be far from wasted. For one thing, it will make it easier to understand how a Touch-Tone® telephone functions. Perhaps even more importantly, many of the projects throughout this book are designed to simulate some function of a rotary telephone. To understand how these projects work, therefore, you obviously must understand how the rotary telephone works.

Then I will introduce the Touch-Tone® system, which is a more complex but more efficient variation on the old rotary dial system. Certain things are easy to accomplish with a Touch-Tone® system that would have been difficult, impractical, or even impossible using the old rotary dial system.

This chapter continues with a fairly extensive look inside a modern electronic telephone. Again, I will be simulating many of the functions encountered here in the various projects throughout this book.

Basic telephone connections

Standard telephones are connected to the telephone lines provided by the telephone company with a cable made up of four wires, even though only two wires extend between the telephone installation and the telephone company's central switching office. To maximize standardization, each wire is assigned a specific color. The four colors used for telephone connections are red, green, yellow, and black.

In most modern telephones, the yellow and black wires are rarely used. Often the yellow wire is shorted to the green lead within the wall jack. This activates the telephone's internal ringer circuitry. As you will see, the projects in this book use just two of the four available wires—red and green. In most modern telephone systems, the black wire is generally left unused, except as a spare wire. Some telephone cables include a fifth wire, usually colored white. Again, this is normally just an unused spare, although the black and white wires might be employed in some multiline systems, such as an office with more than one telephone number.

The yellow wire's primary function is to carry the ringer signal. In most home telephone hookups, the yellow wire is shorted to the green wire within the jack. To disable the telephone's ringer, but otherwise permit full operation of the telephone, just lift the yellow wire and leave it disconnected.

Until recently, home telephone installations were either permanently wired into a jack, or a four-prong jack like the one illustrated in Fig. 2-1 was used. Figure 2-2 illustrates the wiring within this jack.

This old four-prong telephone jack was easy to use. All you needed was a screwdriver to connect it to a cable from a telephone or a related device or project. Loosen the appropriate screw, strip off an inch or two of insulation from the wire, and wrap the bare end around the screw a couple of times. Then tighten the screw to hold the wire firmly in place. Once this sim-

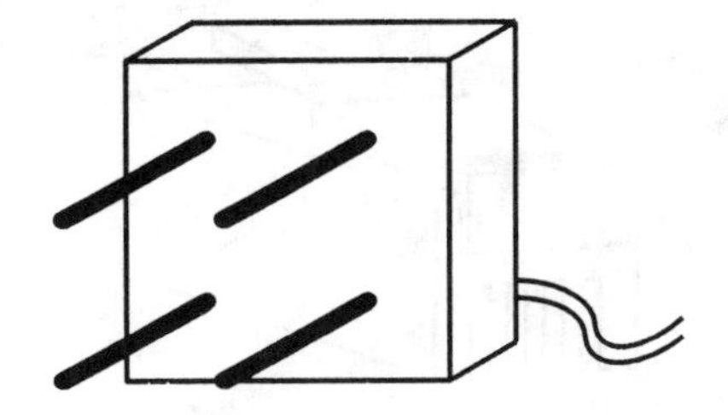

Fig. 2-1 *The four-prong jack used to be standard for nonpermanent telephone installations.*

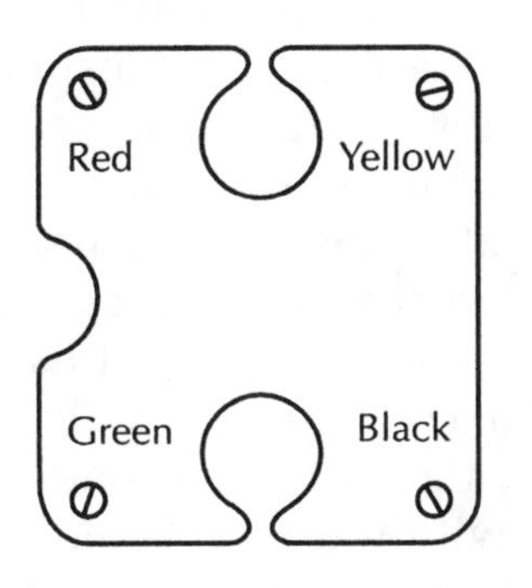

Fig. 2-2 *The standard color-coded wiring in the four-prong telephone jack.*

ple procedure was repeated for each of the required wires (usually just two, but no more than four), the jack was completely connected—that is all there was to it.

While you might see some jacks of this type still in use in a few older installations, they are considered obsolete today. Shortly after deregulation, AT&T came up with a new, supposedly improved, modular jack. This device is illustrated in Fig. 2-3.

The chief advantage of the modular jack is that it is considerably smaller than the old four-prong telephone jack. Unfortunately the smaller size makes it somewhat more difficult to work with. Many prewired cables and adapters are available, but it is tricky to connect bare wires to a modular jack. The wires are very small and must be precisely placed in a very small space. To ensure a good, solid connection that won't pull out, you need to use a specialized crimping tool.

Personally I suspect the real "advantage" of the new, "improved" modular jack is that AT&T owns the patent on it, and the old four-prong telephone jack is in the public domain. Any other company that wants to manufacture any product using a

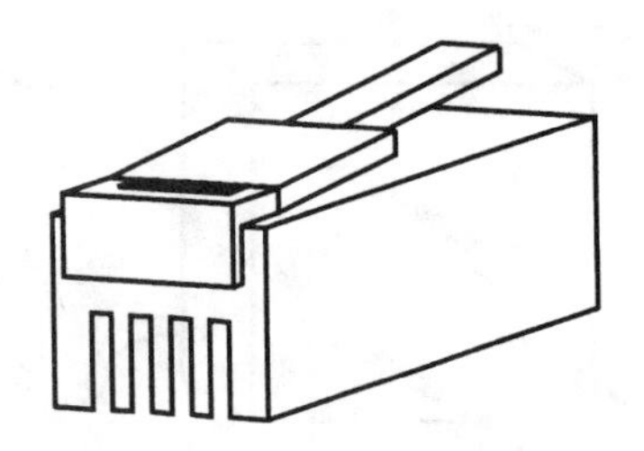

Fig. 2-3 *The new standard for connecting telephone equipment is the modular jack.*

modular jack must pay a small royalty to AT&T. This inevitably increases the overall price to the consumer. Nevertheless, the modular jack is now the standard you have to work with unless you want to make major renovations to all of your telephone-related equipment and replace the telephone company's jacks.

Perhaps the easiest way to use modular jacks with the telephone projects presented throughout this book is not to bother wiring jacks onto bare cables at all. You can buy a premade telephone cable and cut the jack off of one end. Strip each of the necessary the wires within the cable and connect them to your circuit board. Some telephone cables are available with a modular plug on one end and bare wires on the other.

Another approach is to install a modular jack in the housing for your project. It is relatively easy to connect the wires from the project to the rear of the jack. Then you can plug one end of a commercial modular telephone cord into the jack and plug the other end into the existing telephone jack on the wall.

You probably won't have enough wall jacks. If you are just adding one project, you can use a Y-adapter, which permits you to insert two modular plugs into a single modular jack. Alternatively, you can add more jacks. Connect the wires to the rear of each jack in parallel with one another.

Figure 2-4 shows a one-to-four modular jack connection box. Connect a short wire from the single jack in the back to the wall jack. Then you can connect your projects to the four jacks in front. You can easily expand the system for more than four jacks. Just watch out that you don't get too carried away because the ringer equivalence of all devices used simultaneously is cumulative. If you try to use too many add-on projects simultaneously, you could overload your telephone lines and then nothing will work properly. If adding a new project makes your telephone lines function unreliably or erratically, try unplugging something. In most cases, the first thing to be affected by over-

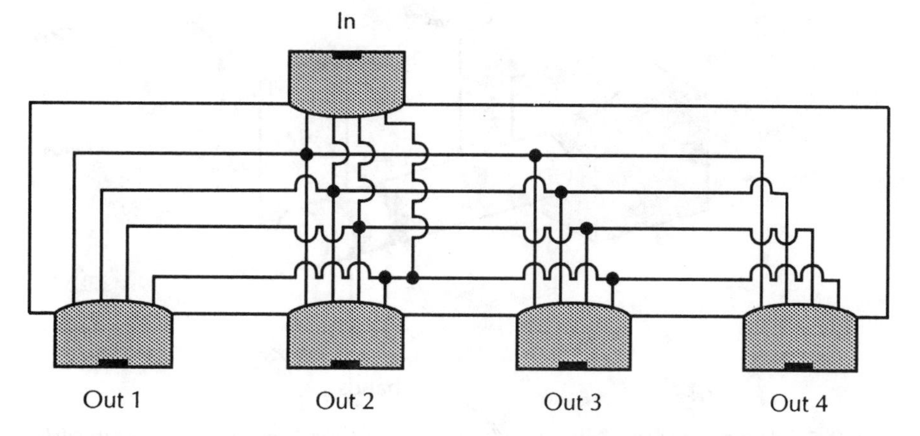

Fig. 2-4 *A one-to-four modular jack connection box permits the connection of multiple telephones (or related devices) to a single standard jack.*

loading is the ringer signals. Your telephone might ring very weakly or not at all, while the voice and dialing circuits work normally.

Telephone signals

Basically just two wires connect your telephone to the central switching office of the local telephone company. These wires are commonly called *tip* and *ring*. It is important to realize that the name ring in this case has nothing to do with the signal used to ring the telephone. These names are throwbacks to the early days of manual telephones and manual switchboards. Back then, when you wanted to place a call, you rang up the local telephone operator and told her who you were trying to call. She then took a wire corresponding to their telephone and manually connected it to a socket corresponding to your telephone, making the electrical connection between the two telephones.

You still might encounter the multiconductor plugs that were designed for early telephone switchboards—the familar phone plug, as shown in Fig. 2-5. There are actually three independent conductors in this type of plug. Notice the pair of small, insulating rings on the metal plug. These permit a three-conductor connection. One connection is made to the tip of the plug, and a second connection is made to the section of the metal plug between the insulating rings—thus, the two connecting wires in the telephone system were logically dubbed tip and

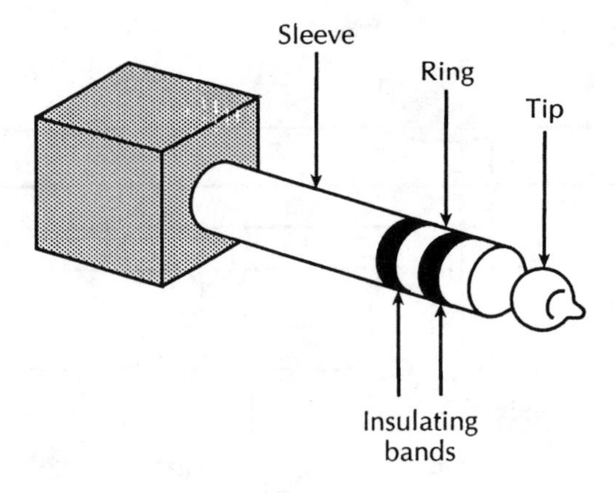

Fig. 2-5 *The terms "ring" and "tip" came from the phone plugs used in old-fashioned telephone switchboards.*

ring. The third connection is made to the sleeve below the insulating rings. This was used for the switchboard's internal grounding. This connection is not of any particular interest to us.

Of course, all telephone service has long since been adapted to automatic switching, and old-fashioned phone-plug switchboards are quaint antiques. But for convenience, the old names were retained. Because these names are just for consistent, easy reference between technicians, there is no point in updating the names. It would just make things confusing.

The telephone company's central switching office serves a number of functions. It supplies the direct current (dc) operating voltage your telephone needs to operate. This is an independent power source, not the output of your local power company, which supplies your household ac power. This is why telephones will usually work even in a power failure, and conversely, why the telephone lines might be out yet your lights and other electrical equipment continue to work just fine.

The central switching office also decodes dial pulses or tones from your telephone when you call out. The telephone company's equipment automatically determines the number you have dialed and makes the necessary connections. Typically a central office can handle up to 10,000 independent telephone numbers. If the number dialed is not in that office's exchange area, the call is automatically transferred to the appropriate central office for processing.

Once the dialed number is identified and connected, the central office transmits a special ring signal to the caller's telephone, so the person will know there is a call. At the same time a ring-back signal is sent back to your telephone, so you know your call is going through. The ring signal sent to the number you are calling is not necessarily synchronized with the ring-back signal you hear. Electrically the two signals are entirely different. The ringer signal is a relatively high alternating current (ac) voltage, while the ring-back signal returned to the caller's telephone is just a normal audio signal, just like the voices during the conversation. This is why sometimes you might hear four rings, while the person you called insists the telephone rang only three times, or vice versa.

The primary purpose of the telephone lines, of course, is to transmit audio frequency voice signals from one telephone to another, permitting conversations at a distance. This is implied by the very name *telephone*. *Tele* essentially means distant, or operating at a distance, and *phone* means sound. So a telephone is a device for transmitting sound over a distance.

Any analog audio signal is an ac signal. In a telephone system, this ac audio signal rides on the normal dc off-hook voltage. The human ear can detect frequencies ranging from about 20 Hertz (Hz) up to 20,000 Hz, but we don't need this full audible bandwidth to transmit spoken voice signals. Most speech signals lie in the 200- to 4000-Hz band. Anything below 200 Hz or above 4000 Hz isn't really needed for a telephone system. Practical telephone systems are designed to restrict the actual transmitted audio signals to an even tighter frequency range than this. High fidelity audio reproduction is not needed for ordinary telephone conversations. The wider the transmitted bandwidth, the greater the odds of random noise getting into the system. The audio passband for telephones is normally restricted to the 300-Hz to 3000-Hz band. This is why your friends' voices don't sound very natural on the telephone. And when somebody tries to send music over the telephone, it rarely sounds as good as the cheapest pocket radio. But voice signals restricted to this frequency range are perfectly understandable, and that is the primary consideration in a telephone system. By restricting the frequency range of the transmitted audio signals, noise interference can be kept to a minimum, and signals can be more effectively and economically multiplexed.

Your voice signal is not just sent to the telephone of the person you are talking to. Some of your voice signal is echoed back to your own telephone, with a slight delay. This echo is caused by small mismatches in transmission line impedances. The most significant mismatches tend to occur at hybrid interfaces between a two-wire circuit and a four-wire transmission system. The echo's delay is determined by the distance between the two telephones connected for the conversation. Too much delay can be annoying, or even confusing, and the telephone company deliberately inserts some loss into the lines to limit echo effects.

However, a little bit of echo appears to be a good thing. It seems to reassure the speaker that his signal is getting through and his telephone hasn't gone dead. It can be quite disturbing to converse on a telephone without hearing your own voice echoed back in your earpiece. This type of echo is usually deliberately created within the telephone itself, because true line echo is far less controllable or predictable. You'd get a greater delay on a long-distance call than when you are conversing with your next-door neighbor.

In addition to the actual voice signals, the telephone lines carry a number of additional signals. For example, there is the dc voltage used to power telephones and related equipment and to signal the central office when the telephone is off the hook.

There are also a number of special coded audible tones to alert someone using a telephone of various conditions. A good example is the ring-back signal mentioned earlier. Another example is the dial tone, which lets you know that the telephone line is functioning and open for placing an outgoing call. Each of these tones is actually comprised of two or more discrete frequencies. In addition, many of the tones are turned on and off in a specific pattern to make them more identifiable.

The dial tone is continuous. It is not pulsed on and off like most of the other audio signals used on telephone lines. It is made up of two tones with frequencies of 350 Hz and 440 Hz.

When you dial a number and the telephone is already in use, you get a busy signal made up of a 480-Hz tone mixed with a 620-Hz tone, pulsed on and off one complete cycle per second. That is, the compound tone is turned on for one-half second, then off for one-half second, then the cycle repeats. This gives the distinctive, familiar "bzzt, bzzt, bzzt" effect.

The normal ring-back signal is made up of 440-Hz and 480-Hz tones that are pulsed on for 2 seconds, separated by a 4-second gap, or tone-off period. For PBX ring-back, the same frequencies are used, but the on-off pattern is altered. In this case the tone is turned on for 1 second and off for 3 seconds.

If you misdial and reach a nonexistent telephone number,

you are alerted by a continuous tone that is frequency modulated at a 1-Hz rate running from 200 Hz up to 400 Hz.

If you leave your telephone receiver off the hook for several minutes without dialing, the telephone company assumes something is wrong. Probably the receiver got knocked off the hook or forgotten. Of course, you can receive no calls under this condition. To alert you that your receiver is off the hook, the telephone company sends a very loud audio signal comprised of four nonharmonic frequencies—1400 Hz, 2060 Hz, 2450 Hz, and 2600 Hz. This distinctive complex tone is pulsed on and off in a 0.1-second on, 0.1-second off pattern. This makes for a very distinctive and hard to ignore signal. Unfortunately this means you can't just leave the telephone receiver off the hook when you don't want to be disturbed by an incoming telephone call. Perhaps you might want to take a nap, and don't want to be disturbed by a ringing telephone. But after just a few minutes (the exact time varies from area to area and from telephone company to telephone company) you get the off-hook alert signal. If you have a four-prong or modular jack on your telephone, you can simply unplug it. Instead of getting a busy signal, anyone attempting to call will hear a regular ring-back signal, indicating an unanswered telephone. Many modern telephones have a switch that permits you to turn off the ringer, which has the same effect as unplugging your phone as far as any attempted caller is concerned.

The various standard audio signals used in United States telephone systems are summarized in Table 2-1.

The telephone lines also carry a number of inaudible control signals that are transparent to callers on the line. These control signals are used to aid in making the necessary interconnections between central offices, and for other in-house purposes. Some of these control signals are analog, while others are in digital (on-off) form. For the purposes of this book, you really don't have to concern yourself with these internal control signals.

Placing a call

In most telephones, when the handset is on its hook, its weight holds down one or two normally closed spring-loaded buttons, holding them open. This breaks the dc circuit between the telephone and the central switching office. The ac ringer circuit remains connected. A large capacitor blocks any dc voltage

Table 2-1 Typical audio frequency tones used in telephone systems.

Dial tone	350 Hz + 440 Hz	Continuous
Busy signal	480 Hz + 620 Hz	Pulsed—0.5 second on, 0.5 second off
Ring-back	440 Hz + 480 Hz	Pulsed— 2 seconds on, 4 seconds off
PBX		Pulsed—1 second on, 3 seconds off
Receiver off hook	1400 Hz + 2060 Hz + 2450 Hz + 2600 Hz	Pulsed— 0.1 second on, 0.1 second off
No such number	200 Hz to 400 Hz	Continuous Frequency modulated sweep: sweep frequency = 1 Hz

components in this signal. When the central office sends a ringer signal to your telephone, the bell (or electronic sounder) alerts you to the incoming call. The ringer circuit has no effect on the voice signals transmitted back and forth during your conversation because this ac circuitry is designed to present a very high impedance to the audio signals. The ringer signal is blocked when the telephone is off the hook. It can only get through when the telephone's handset is on the hook.

Removing the handset releases the spring-loaded cradle buttons. They return to their normally closed position, completing the circuit between your telephone and the central switching office. A dc current now flows through the circuit. If the central office sends you a ringer signal to complete a call placed by someone else, an audio connection is made between the caller's telephone and your instrument. The conversation can now proceed. When one party hangs up their handset, the cradle switches are reopened, and the circuit between the two telephones is broken. The call is terminated.

What happens when you pick up your telephone handset when no ring signal has been sent to your line? The central office sends you a dial tone to let you know that your telephone line is live and functional, then it waits for you to begin dialing the number you want to call. In most modern telephone systems, there is also a timer that reacts if your telephone is off-hook for more than a preset period of time without a call being placed. The telephone company assumes your receiver has been knocked off the hook (my cat is notorious for this), or that it has been set down and forgotten. After the timer's waiting period is over, the central office sends a loud alert signal to your telephone. It is loud enough to hear a good distance from the earpiece. This signal alerts you to hang up your telephone so you can receive any incoming calls.

Taking the phone off the hook is not a practical way to avoid being disturbed by unwanted calls. Most modern telephones use a modular plug that can be easily disconnected. Anyone calling you will hear the regular ring-back signal, as if your telephone was ringing but not being answered. Many modern telephones feature a switch that permits you to silence the ringer without physically disconnecting the entire unit.

When you pick up your handset and begin dialing, the dial tone stops. The central switching office decodes the number you dial in and makes the appropriate connections, either to another telephone within the same exchange or to a second central office in a different exchange, which connects you to the appropriate line.

There are two basic types of dialing used in modern telephone systems. One is the pulse dialing method, used in older rotary dial telephones. The more modern approach is the dual-tone multifrequency (DTMF) system. This is often called the Touch-Tone® system. Actually "dialing" isn't strictly the correct term to use for DTMF systems, because there is no dial. The number is entered via a matrix of push-button switches.

It is important to realize that not all push-button telephones are DTMF devices. Some electronically simulate pulse dialing. As far as the central switching office is concerned, it is exactly the same as using a rotary dial telephone. Many commercial push-button telephones today include a switch to select between pulse and DTMF dialing.

Virtually all telephone networks today will accept pulse dialing. Most, though not all, modern telephone networks will also accept DTMF dialing. The DTMF system is rapidly becoming the standard, but it is still not available in all areas, particularly in rural or sparsely populated communities. Remember, you can use pulse dialing on a DTMF network, although the reverse might not be true.

Pulse dialing

Pulse dialing is relatively simple to accomplish either electromechanically (as in old-fashioned telephones) or electronically (as in newer telephone equipment). A typical dial pulse signal is

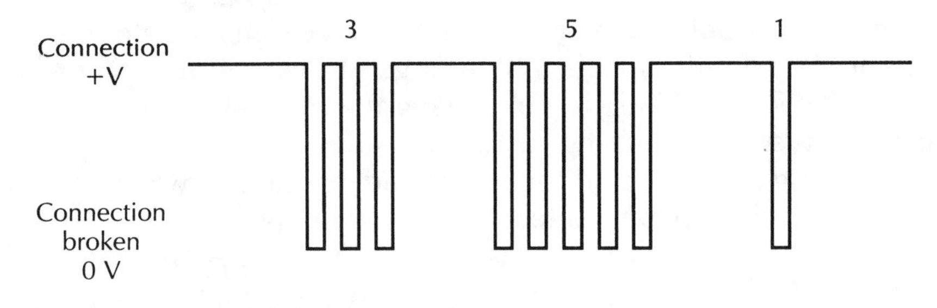

Fig. 2-6 A typical dial-pulse signal.

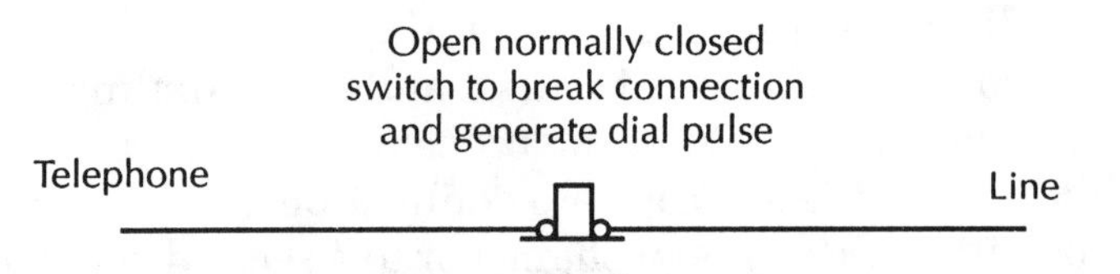

Fig. 2-7 Dial-pulse signals can be generated with a simple SPST switch.

illustrated in Fig. 2-6. As you can see, this is just a pseudo-digital signal, made up of high and low levels with (ideally) no transition time between them. The signal is always either fully high or fully low, and never at any intermediate level. If you assume the high level is some positive voltage (let's call it V+), and low level is at ground potential (0 volts (V)), you can manually generate signals of this type with a simple single-pole, single-throw (SPST) switch, as illustrated in Fig. 2-7. This switch can be either a mechanical switch or an electronic switching circuit. The effect is the same in either case.

This is what happens in a pulse-dialing telephone. The connection from the telephone's voice circuits to the line to the central switching office are opened (Low) and closed (High) in a specific pattern to indicate the specific number being dialed. The patterns are quite simple and obvious. If there is just a single pulse, then the number 1 was dialed. If you dial 2 you'll get two pulses, if you dial 3 you'll get three pulses, and so forth. This continues for each of the ten available digits from 1 to 0 (in telephone usage, 0 counts as 10).

Theoretically you could "dial" each digit by pressing and releasing the telephone's cradle (hang-up) buttons in the desired pattern. However, besides being awkward and inconvenient, the required timing between pulses is critical. If there is too great a pause between adjacent pulses, the central switching office will think a new digit is being entered. For example, if you want to dial 5, but pause too long between the second and third pulses, the central switching office will assume you meant to dial 2 and then 3. It would take a great deal of practice to dial a number correctly in this fashion, and frankly I can't imagine why on earth anybody would ever bother.

The classic solution was the rotary dial, as illustrated in Fig. 2-8. As the dial rotates, it makes and breaks the appropriate number of connections. Notice that the dial has to travel further for 6 than for 2 to pass through the correct number of contact points to create the desired number of pulses. A mechanical stop is included to prevent the dial from being turned past 0 (10). As you move the dial forward, nothing happens electrically. The dial is spring-loaded, so it automatically returns to its normal rest position (above 1). While the dial is moving backwards, the switch contacts generate the pulses. The physical spacing between the contact points on the underside of the dial determines the spacing of the individual pulses within a single digit. Because you

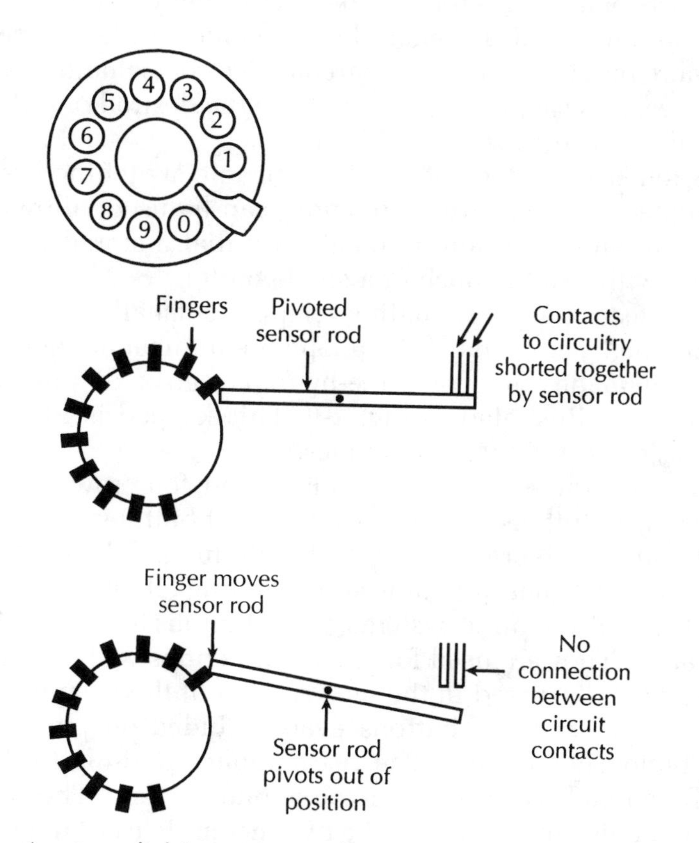

Fig. 2-8 A rotary dial is used to make and break the connection for the appropriate number of dial pulses.

have to wait until the dial returns to its rest position and then rotate it forward before the pulses for the next digit begin, there is always a sufficient gap between digits so the central switching office can readily distinguish between them. If you dial 4 and then 2, there is no way the central switching office will think you dialed just a single 6.

The pulse rate is 10 Hz, or 10 open/closed cycles per second. Because the pulse rate is fixed, but the number of pulses varies, higher digits (such as 9 or 0) take longer to transmit than lower digits (such as 1 or 2).

When telephones were first developed, the field of electronics as we know it didn't really exist, except as a laboratory curiousity, so an electromechanical system (the rotary dial) was the best available approach to generate the dial pulses. But any system with mechanical moving parts is subject to physical wear and tear, dirt accumulation, and mechanical problems such as bent parts, jamming, and the like. With today's technology, an all-electronic circuit is generally more efficient, less expensive, and more reliable than a comparable electromechanical system. Of course, it is easy enough to generate the desired on-off pulses with an electronic switching circuit.

Another advantage of an all-electronic switching system is that some sort of electronic memory can be used, allowing the circuit to remember and automatically dial a stored telephone number with just the push of a single button.

An electronic pulse-dialing telephone usually doesn't have a rotary dial. The digits of the telephone number are entered via a keypad made up of a three-by-four matrix of push-button switches, as illustrated in Fig. 2-9. This keypad is identical to the ones used in DTMF telephones.

Because there are three columns and four rows, there are twelve push buttons, so there are ten digit buttons—from 1 to 0 (10)—and then there are two "extra" buttons, labeled * and #. These spare buttons are not used for dialing telephone numbers. In fact, in pulse-dialing systems, these buttons are rarely, if ever, used at all. They are used for special functions with DTMF dialing, and are discussed in the next section of this chapter.

So why are these buttons even included on pulse-dialing push-button telephones? The reasons have nothing to do with any functional aspect of the telephone system. They are included for visual symmetry. The two incomplete columns would look funny. Also, it is cheaper for manufacturers to use the same keypads for both pulse and DTMF telephones (not to mention

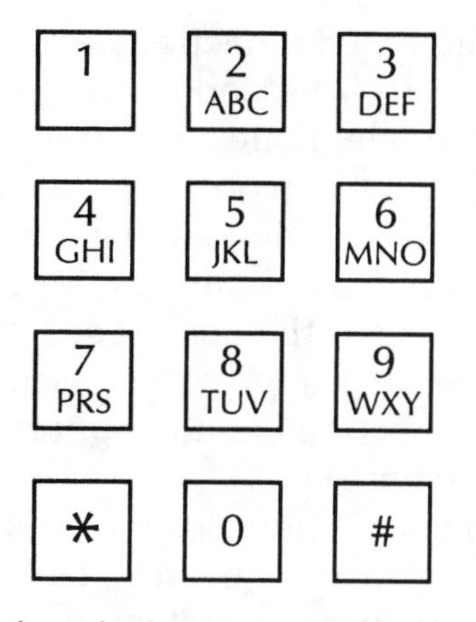

Fig. 2-9 *The digits of the telephone number can be entered via a keypad made up of a three-by-four matrix of push-button switches.*

telephones with a switch to select between the two dialing systems). I've even seen a few pulse-only push-button telephones in which these two buttons were not internally connected to anything within the instrument.

Incidentally, it seems rather inappropriate to dial a number on a push-button telephone, because there is no dial. This is another case where the old term has been held over simply for convenience (like the ring and tip wires). So we end up dialing without a dial.

Connecting with the central switching office

The telephone company owns the wiring up to a terminal unit or terminal block, which is usually either in the basement or in a secured box on the outside wall of a building. From that point on, the wiring is the responsibility of the building's owners and/or residents.

The terminal block might be referred to by different names, depending on the local telephone company's preference. Some common names for the terminal unit are demarcation point arrangement (DPA), network interface (NI), and standard network interface (SNI). There is really no practical difference between these names. You might encounter additional names for the same thing.

The terminal block is nothing complicated or difficult to understand. It is just a small box in which the telephone lines are terminated. It is simply a connection box, permitting your individual telephone lines to be connected to the main telephone lines provided by the telephone company. If you open up a telephone terminal block, you won't see any fancy, high-tech circuitry—just some wires connected to a few binding posts.

Inside the terminal block are three connection posts. The center post is connected to the nearest secure earth ground, such as a cold water pipe. A heavy wire is used for this connection. Most other local connections in a telephone system usually employ 22- or 24-gauge wire.

The outer two connection posts connect to the actual telephone lines, running to the nearest central switching office. These connections are fused to protect the telephone lines if they are struck by lightning or a fallen high-voltage power line. When these fuses blow, the potentially dangerous current is diverted to ground, minimizing the possibility of it starting a fire or damaging equipment.

The connection to the telephone company's central switching office is made via two wires. All of the telephones connected to the same central switching office are called an *exchange*. If you call a number in a different exchange, the call must be routed to another central switching office. The first three digits of the standard seven-digit telephone number identify the exchange, or the appropriate central switching office to make the connection. You might have noticed that most telephone numbers given in fictional television programs use the 555 exchange. This prevents the possibility of accidentally duplicating an existing working telephone number. The 555 exchange is not an operating exchange anywhere in the country. This partical exchange designation is reserved for special testing purposes within the telephone company itself. There is no law requiring television programs to use the 555 exchange for fictitious telephone numbers, but it saves their producers a lot of hassles, and even invasion of privacy lawsuits.

Exchanges are grouped into *areas*, identified by an extra three-digit area code. Usually you must first dial a 1 or 0 before dialing the area code of a long-distance telephone number. This helps prevent any possible ambiguity as far as the central switching office is concerned when it attempts to make the appropriate connections. All United States area codes have either a 0 or a 1 as the middle digit. The first digit is never a 0 or 1.

Because three digits are used to identify the exchange, it would seem that any given area code could handle up to 1000 exchanges (from 000 to 111). But not all possible three-digit exchange numbers are used. As already mentioned, the 555 exchange is never used for a practical exchange. Exchange numbers never begin with 0 or 1, as this could cause confusion with long-distance dialing. There are a few other minor restrictions, but there are still plenty of available exchange numbers to choose from. The greater the population density, the faster the available exchange numbers are used up. This is why the state of New York uses at least eight area codes (and this number can be increased as needed), while all of the state of New Mexico is covered by the single 505 area code.

A basic (local) telephone number consists of seven digits. The first three digits identify the exchange, as discussed previously. This leaves four digits to identify the individual connections within a given exchange. There are no universal restrictions on the final four digits of a telephone number, so a given exchange can handle up to 10,000 customers, with numbers running from 0000 up to 9999. Most exchanges do not use all of the available numbers, leaving some headroom in the system for new customers and unforeseen circumstances.

When you dial a number, the digits you dial are sent through the two wires in your terminal block to the central switching office for your exchange. If the number you dial has a different exchange, the call is automatically switched to the appropriate central switching office for that exchange. Then the central switching office decodes the final four digits to identify the party you are trying to contact, making the suitable electrical connections to complete the circuit between your telephone and the telephone belonging to the person you are calling.

The earliest telephone systems used live operators to manually make and break all connections via patch cords on a switchboard. This worked in a crude way, but as telephone service became more popular and widespread, the limitations and inefficiency of such a system quickly became very apparent. To provide widespread telephone service there needed to be some way to automate the process of making the connections to complete calls. The first significant automation system was called *Strowger switching,* named after its inventor. This system was patented in 1891, but it was not adopted by Bell Systems until 1918, and it quickly became the norm. By the mid-1970s, fully half of all Bell System exchanges used Strowger switching.

Strowger switching is also sometimes called *step-by-step switching,* which is a more descriptive name. Several switches in series (a "switch train") work together in progressive step-by-step operation to connect a caller with the telephone number they want to reach. Each step in the process is directly controlled by the dialing pulses described earlier. Electromagnetic relays are used as the actual switching devices.

When someone lifts their telephone handset off its hook, current flows in a loop between their telephone and the central switching office. This current flow is sensed by a relay that activates the first switch in the train. This first switch is called the *linefinder.* It steps vertically until it finds the connection to the off-hook line. Then it steps horizontally to locate the first idle selector switch. Any selector switch that is presently in use completing some other call circuit is ignored and jumped over.

When a free selector is located, a dial tone is sent back to the caller's telephone. The system is now ready to accept and process the dial pulses as the number is dialed. If the caller begins dialing before hearing the dial tone, the system can become confused, and the caller might not be able to reach the party being called without hanging up and starting over. Fortunately, unless the central switching office is very small and/or very busy, these initial switching steps take place very quickly. Usually you won't even notice any delay at all. The dial tone will seem like it was sounding the very instant you lifted the handset from its cradle. Actually it was turned on in the time it took you to move the receiver from the cradle position to your ear.

The first selector switch waits for the first digit of the telephone number to be dialed, then it steps vertically one step for each dial pulse it detects. For example, if the first dialed digit is 2, the first selector switch moves over two positions.

Once the first selector has repositioned itself in response to the dialed digit, the process is repeated with a second selector switch for the second dialed digit. This sequence is repeated for each digit in the telephone number. At the end, your telephone line is connected to the line leading to the telephone of the person you are calling. The central switching office sends a ring signal out to the called phone (and a ring-back signal to the caller's telephone) until either the called party answers their telephone (the voice circuit is then completed), or the caller hangs up (the connection is completely disconnected and all of the selector switches are freed for another call).

If the telephone for the number you have dialed is already off the hook, the connection cannot be completed. You will hear a pulsed busy signal to let you know what has happened.

If all of the linefinder switches are busy with other calls when you try to place a call, you won't be able to get through. In the original Strowger system, the caller would not get a dial tone, so there was no point in attempting to dial the number. Today's telephone systems are more complex, and the dial tone might be heard even when it is impossible to place a call. When you attempt to dial, you hear an alert tone and a recorded message telling you that all the lines are busy at the moment. Most modern central switching offices are large enough and efficient enough that this rarely happens.

The Strowger switching system is roughly illustrated in Fig. 2-10. This is not intended as an exact schematic diagram, of course, just an illustration of the basic operating principles involved in the system. The Strowger switching system required a large number of large electromechanical relays. In any system, the more moving parts there are, the greater the chance of something getting stuck or bent. The mechanical reliability of the individual switches was low, so trained maintenance personnel were required to be at the central switching office at all times. This added to operating costs.

The working environment at the central switching office was certainly less than pleasant, with all those large relays constantly clicking back and forth. The Strowger system inevitably generated quite a bit of mechanical and electrical noise. It could also cause interference to nearby radio receivers.

Another significant limitation of the Strowger switching system was that everything was hard-wired together. Therefore, it wasn't easy or convenient to make changes in the switching arrangement.

The hefty maintenance costs, inflexibility, and other limiting problems of the Strowger switching system soon forced Ma Bell to make some major improvements and refinements in the system. To correct for the limitations, a new type of switching matrix was developed, using common control. This new type of switching matrix was dubbed the *crossbar.*

The common control element is a form of digital computer made up of numerous relays. It is programmable to a limited extent. It stores the dialed number and feeds it out to the switching matrix according to a number of predetermined rules. Some of these rules are hard-wired into the circuitry, while others are programmable. The rules allow for anticipated variations in placing local and long-distance calls. They also tell the system how to choose an alternate route in case the originally chosen

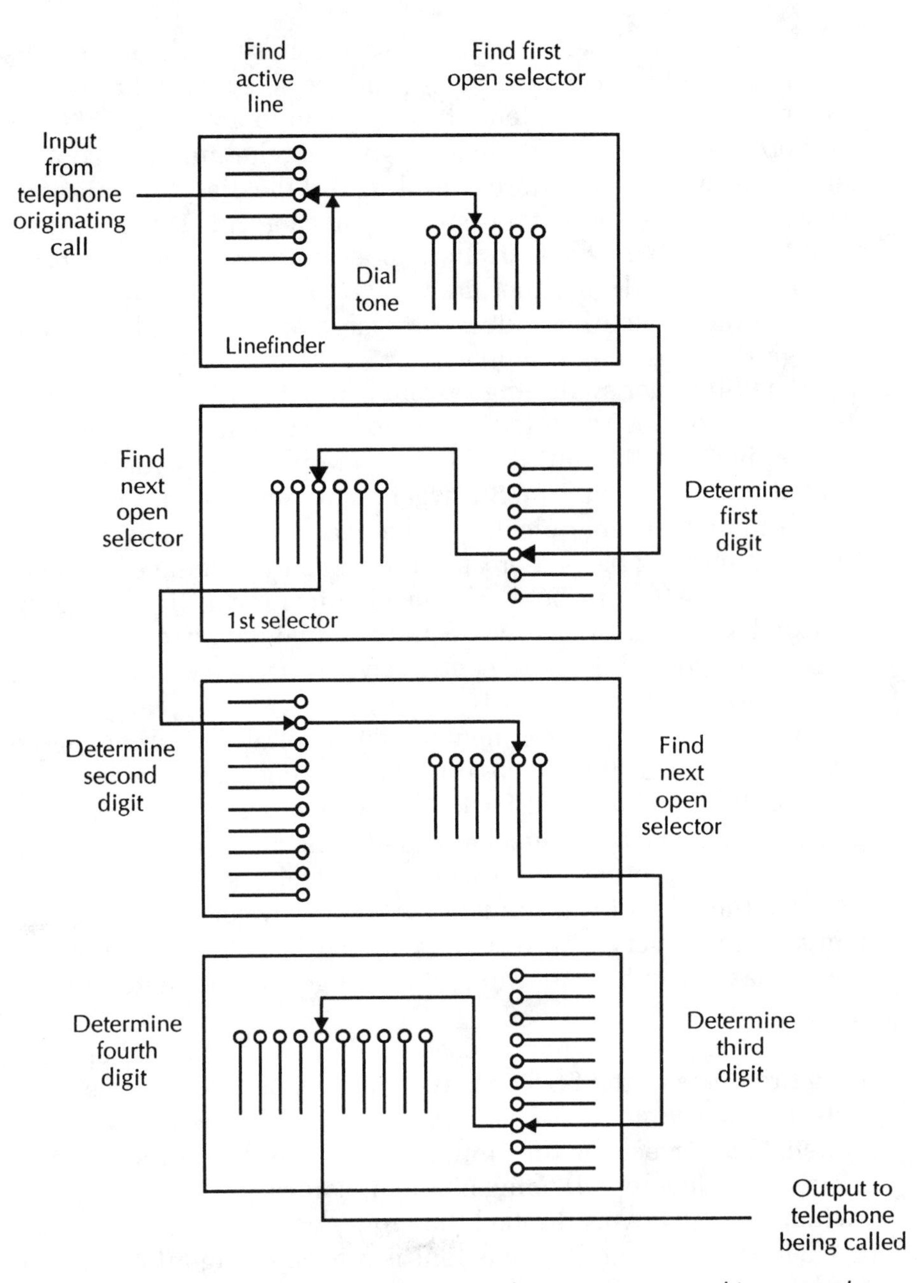

Fig. 2-10 *Some form of the Strowger switching system is used in most telephone central switching offices.*

route is busy. If the call is somehow blocked or encounters a fault in the switching path, the rules tell the system how to reroute the call. Because the dialed number is stored in the pro-

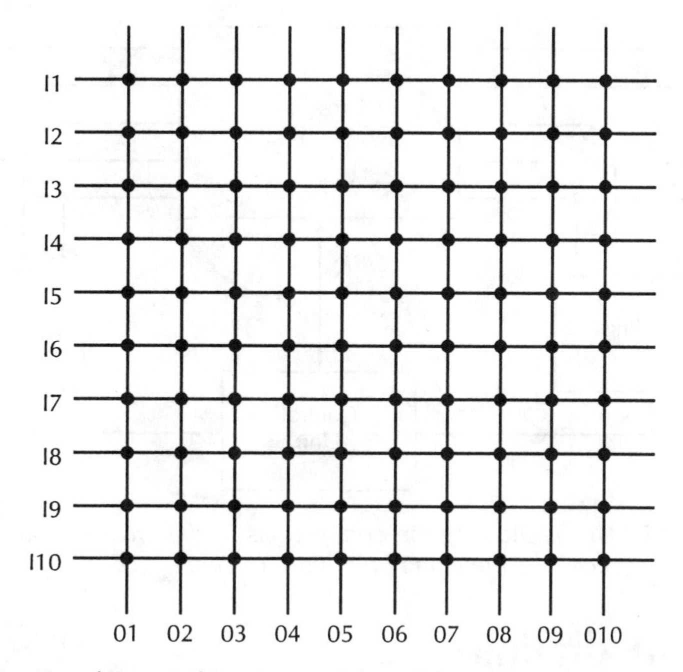

Fig. 2-11 *Crossbar switching is a useful modification on the basic Strowger switching system.*

grammable computer relays, it can be recalled for reuse so the routing can be started over from the beginning, if necessary.

The crossbar system is simply a matrix network, as roughly illustrated in Fig. 2-11. In operation, one vertical line and one horizontal line are activated, and the connection is made at the point where they cross each other. If the vertical lines are used for input signals and the horizontal lines for output paths, you can connect any input to any of the available output paths. Figure 2-12 illustrates how the connections in the crossbar matrix are directed by the common control computer. To a large extent, the crossbar switching matrix is a refined variation on the basic Strowger system discussed previously.

Increasingly central switching offices are being converted over to all-electronic solid-state switches, which are considerably quieter, smaller, and more reliable than electromechanical relays. Such electronic systems can respond to either simple switching pulses, like those dealt with here, or audio frequency tones, used in DTMF dialing. This is discussed in the following sections of this chapter.

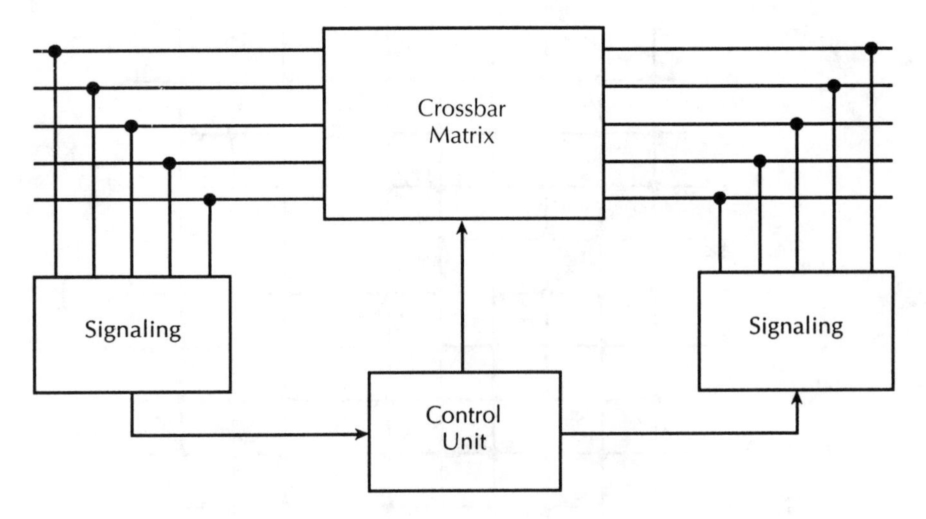

Fig. 2-12 *This is the way the connections in the crossbar matrix are directed by the common control computer.*

DTMF dialing

Dual-tone multifrequency (DTMF) dialing is increasingly becoming the norm in modern telephone systems. This system is often called the Touch-Tone® system. The system was developed in the 1960s, but didn't really catch on with the general public until the mid-1970s. Nowadays DTMF dialing is becoming the accepted standard. It is replacing the old pulse-type dialing system for most popular telephone-related applications.

In DTMF dialing, there isn't a discrete number of on-off pulses. Instead, each dialed digit is represented by a combination of two enharmonic tone frequencies. *Enharmonic* means that the two frequencies are selected so one is not an exact whole-number multiple of the other. For example, 500 Hz and 750 Hz are harmonics of 250 Hz ($2 \times 250 = 500$; $3 \times 250 = 750$), but 333 Hz is not—250 Hz and 333 Hz are enharmonics.

Why are two tones used for each digit instead of just one? For better reliability. If just a single tone frequency is used, the system could conceivably be confused by a random noise signal that just happened to be at the right frequency. But the odds are astronomically against two exactly right frequencies simultaneously occurring because of random noise.

All DTMF dialing telephones use the standard push-button keypad. There are four rows and three columns for a total of twelve buttons. This accounts for the ten numerical digits (1 through 0), and two extra "special function" buttons marked * and #. These extra buttons are not used for dialing telephone numbers. As far as the telephone company is concerned, these buttons are uncommitted spares.

Many companies use computerized telephone systems for connecting callers to appropriate extensions or to permit placement of orders for shipments of various types. The * and # keys are often used with such computerized systems.

There is often confusion about what to call these two keys. The * key is usually referred to as the *star key* or the *asterisk key,* which is clear enough. But the # key has been given many different names, including *pound key, number sign key, number key, crosshatch key,* and *double-cross key.*

Each row in the keypad matrix activates a specific low-frequency tone, while each column controls a specific high-frequency tone. These two frequencies are combined into the dual-tone multifrequency signal. The four row tone frequencies are 697 Hz, 770 Hz, 852 Hz, and 941 Hz. The three column tone frequencies are 1209 Hz, 1336 Hz, and 1477 Hz.

Notice that all of these frequencies were carefully selected so that none of them are harmonically related to any of the others. That is, all of the DTMF frequencies are enharmonic. This

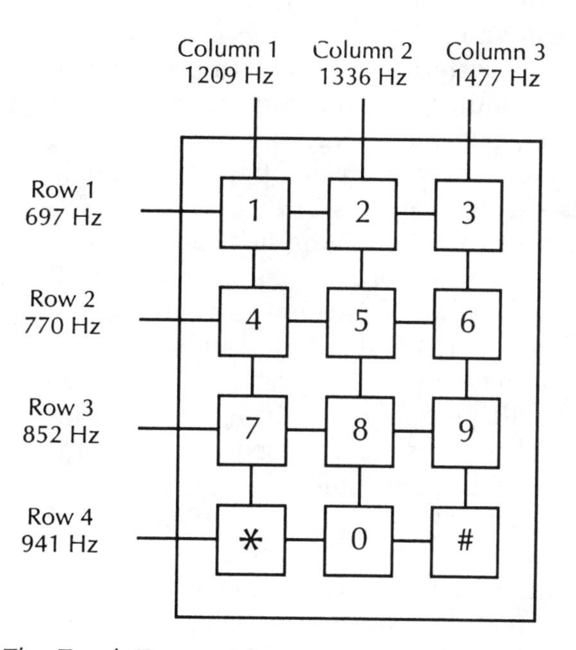

Fig. 2-13 *The Touch-Tone® DTMF system uses discrete frequencies for each row and column on the keypad.*

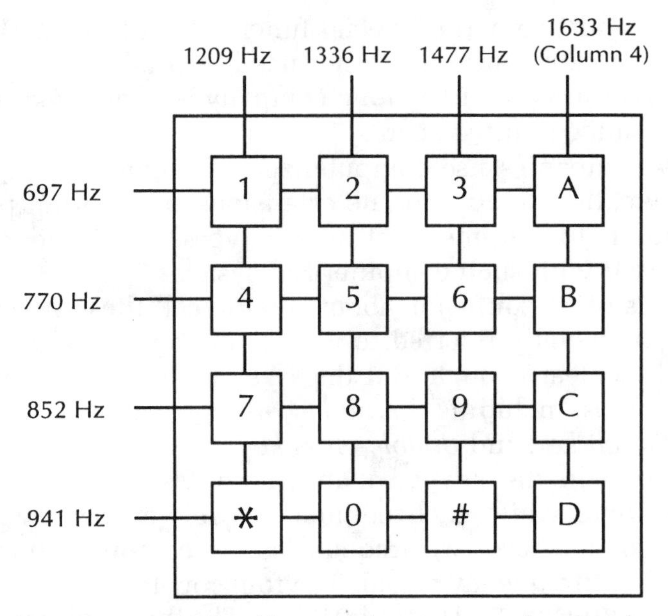

Fig. 2-14 *Some specialized keypads have an extra fourth column for a total of sixteen keys.*

maximizes the efficiency and accuracy of the system, and minimizes the possibility of false triggering because of random noise. The combinations of frequencies for each of the twelve available keys are as shown in Fig. 2-13.

The DTMF system is used for more than just home telephones. Some industrial and computer DTMF systems use a four-by-four keypad matrix instead of the more familiar three-by-four keypad matrix. As a hobbyist, you might occasionally come across surplus keypads of this type. The four-by-four keypad is essentially the same as the smaller, more common version, except an extra fourth column has been added. Its column frequency is 1633 Hz. The four new keys are labeled A, B, C, and D. This type of keypad is illustrated in Fig. 2-14.

If you purchase a four-by-four keypad for use in a telephone project, you will probably simply leave the four extra keys in the added column unused and disconnected in the circuit. They won't do any harm, other than taking up a little extra space on your project's control panel. But if you get a good price on surplus equipment, it is probably worth the minor inelegance of a few nonfunctional keys on your keypad. If you are fussy about such things, you'll just have to purchase an appropriate three-by-four keypad, even if it costs a little more.

Telephone specifications

For obvious reasons, any telephone must be designed to function under a fairly wide range of electrical, mechanical, and acoustical conditions. For example, a telephone that is too fussy about its operating voltage will probably work very erratically, because many factors can vary the actual input voltage seen by the telephone instrument when it is connected to the telephone lines. In practical use, telephones are often subjected to rather rough physical treatment. The instrument must be well designed to stand up to such abuse. As far as acoustics are concerned, it should be possible to carry on a telephone conversation without needing a completely silent environment. In practical use, sometimes a telephone is used when it is quiet and sometimes when the area is quite noisy. Designers of telephone equipment try their best to make such factors as irrelevant as possible, but 100 percent success can never be achieved. If, for example, you have ıe stereo blaring loud enough to make the walls shake, you can hardly blame the telephone if you can't hear what the party on the other end of the line is saying.

There is a lot a variety in modern telephone designs, much more so than a decade or two ago, but some factors are standardized. Some of these factors are standardized by the way human beings are expected to use the instrument. There isn't too much flexibility in designing the overall dimensions of the handset, for example, because the mouthpiece must be near the user's mouth and the earpiece near the ear at the same time. The standard dimensions of the telephone handset are determined largely by the dimensions of the average human head. Similarly, the volume of the reproduced sound at the earpiece must be within certain limits so it is loud enough for most people to hear clearly, without being painfully loud. A possible exception might be a telephone instrument designed specifically for the hard of hearing, which might use a higher output volume. A more practical solution for most hard-of-hearing telephone users is a separate add-on amplifier of some sort. Some are designed to fit over the earpiece and amplify the sound before it reaches the user's ear.

Several parameters in telephone design exist for purely historical reasons. A typical example is the ringing voltage and frequency. For the old-style electromechanical ringer (discussed later in this chapter), a fairly large voltage was required. A modern electronic ringer circuit could just as easily be designed to trigger off of a much lower voltage. But the old standard is still adhered to for compatibility with existing systems.

Of course, some design parameters are defined by the natural characteristics of some of the components used in the telephone instrument. For example, a carbon microphone (or transmitter) requires a certain minimum voltage, and a relay demands a certain level of current before it can operate properly and reliably.

Design of telephone-related projects is made a little easier by the fact that none of the standard operating parameters in the telephone system are terribly critical. For example, the nominal dc operating voltage is 48 V, but the absolute voltage limits run from 47 to 105 V. This possible variation is important. If you use a 50-V or 75-V capacitor in your telephone project, it could be blown if the actual voltage on your telephone line approaches the upper limit. Even a component rated for 100 V is not safe in such an application.

The ring signal is nominally a 20-Hz signal with a voltage of 90-V rms. There is a lot of practical variation in this case. The actual ring signal frequency can drop as low as 16 Hz or go as high as 60 Hz (50 Hz is the absolute maximum ring frequency in some European countries), while the actual ring voltage can be anywhere from a low of 40-V rms up to a high of 130-V rms. Obviously any project connected to the telephone lines must be able to withstand ac voltages up to 130-V rms. I'd up this to at least 150 V to leave a little headroom for safety. Any circuit designed to recognize and respond to the ring signal must be designed to accept the full range of actual ring signals that might appear on any given telephone line.

Nominally the operating current consumed by a typical telephone is given as 20 to 80 milliamperes (mA). Some practical devices might consume currents up to 120 mA. The standardized 20-mA lower limit seems to be a realistic minimum limit, however. Add-on telephone equipment, such as answering machines, amplifiers, and the various projects presented throughout this book, should be designed to remain within similar current consumption limits. The lower the current consumption, the lower the device's contribution to your line's overall load will be. If your project requires a relatively large amount of current to operate, it is best to use a separate power supply, rather than relying on the operating voltage from the telephone lines themselves. You will notice that this is done for a few of the projects presented in later chapters of this book.

The loop resistance of an individual user's line should vary between 0 and 1300 ohms (Ω). In some systems, the actual individual loop resistance might be almost three times as large. The absolute maximum individual loop resistance rating is 3600 Ω. Of course, the individual loop resistance can never drop below 0 Ω. The nominal loss in an individual loop is about 8 decibels (dB), but in some cases this can increase up to 17 dB. Of course, if you add on a lot of extra equipment to the line, the cumulative load can lead to excessive line losses.

If there is excessive distortion of the audible signal the telephone will be rendered useless. If you can't understand what the party on the other end of the line is saying, what good is it? The standard maximum distortion rating for commercial telephones is −50 dB.

❖3 Inside the telephone

Now I can start to get to the "good stuff" as far as the electronics experimenter is concerned. I'll take a look inside the telephone instrument itself, and see what it does and how it works.

The earliest telephones were strictly electromechanical devices. Today most telephones use electronic circuits to perform the same functions and to add new features that were impractical, or impossible, to implement using the older technology.

A general, simplified block diagram of a typical telephone is shown in Fig. 3-1. Any special features, such as automatic redial, require additional circuitry, of course. For now I am only concerned with the basic functional sections required in any practical telephone.

Notice that only the ringer is always connected to the telephone lines. Everything else is disconnected from the lines when the hook switch is held open by the weight of the handset in its cradle. This means all the telephone can do under these conditions is sound its ringer when an appropriate ring signal is received from the telephone lines. All of the other circuitry in the telephone is inactive at this point.

When the handset is lifted from the cradle, the spring-loaded hook buttons return to their normally closed position, completing the circuit between the telephone lines and the rest of the telephone instrument. The telephone company discontinues the ring signal when it detects the off-hook current. The ringer circuit (or device) has a very high impedance to the audio signals transmitted back and forth between the two telephones at either end of the connection, so it is effectively transparent.

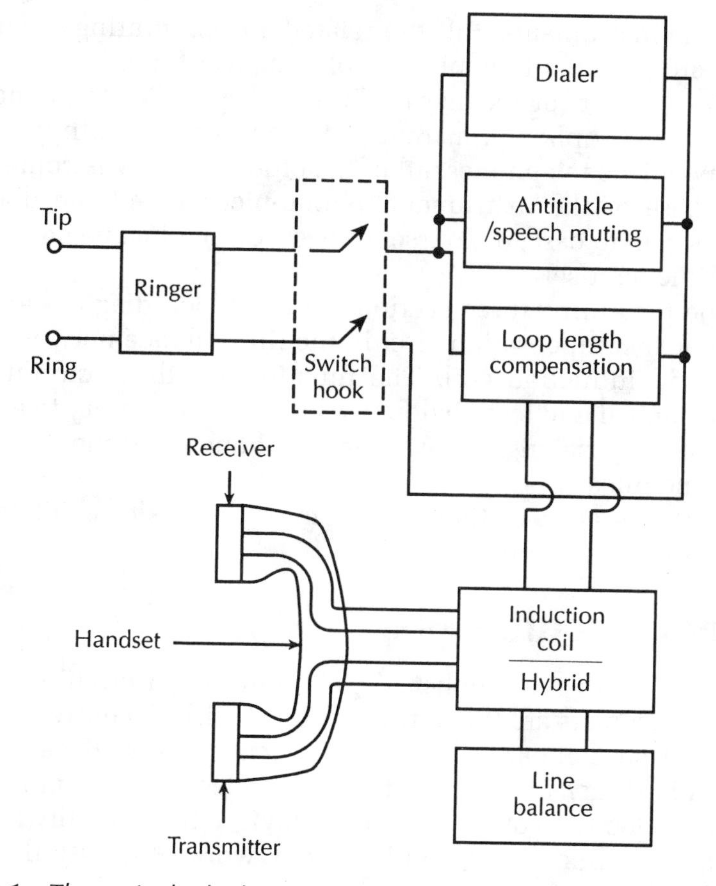

Fig. 3-1 *The typical telephone is made up of several standard sections.*

When there is no ring signal on the line, the ringer inside the telephone effectively doesn't exist. More precisely, it doesn't do anything but sit there.

The dialer can be any of three basic types. It can be an electromechanical rotary dial, an electronic pulse generator circuit, or a Touch-Tone® dual-frequency generator circuit. Some telephones include both an electronic pulse-generator circuit and a DTMF tone-generator circuit, either of which can be selected by a simple manual switch (usually on the bottom of the instrument). When one is switched into the circuit, the other effectively doesn't exist, and vice versa. Both generator circuits are never an active part of the telephone simultaneously—it must be one or the other.

The antitinkle circuit is only found in older electromechanical telephones. It really isn't needed in modern all-electronic telephone instruments. But the related speech muting circuit is a standard part of all telephones, of whatever type.

The speaker (or receiver) is in the earpiece of the handset, while the microphone (or transmitter) is in the mouthpiece. In most two-piece telephones, all of the other circuitry is contained in the base of the instrument. A one-piece telephone doesn't have a separate base, of course, so everything must be contained within the handset.

The remaining three sections in our block diagram are the loop length compensation circuit, the line balance network, and the hybrid induction coil. The functions of these circuits are more technical and less obvious, but they are necessary to ensure the proper interfacing between the telephone instrument and the telephone lines.

I discuss each of these sections of the basic telephone in more detail in the following pages.

Ringer circuits

Chapter 4 looks at a number of electronic ringer circuits. Today electronic ringers are the norm. But until fairly recently, all telephones used standard electromechanical ringers. Because any electronic ringer circuit, by definition, electrically simulates a standard electromechanical ringer device, it is worthwhile to spend some time examining how they work. Actually they are ingeniously simple devices.

First, consider what ringers are designed for. Obviously their purpose is to alert a customer of the telephone company that an incoming call needs their attention. While the telephone is ringing, some of the telephone company's control equipment is tied up, limiting its availability for other calls. Therefore the telephone company wants their customers to answer their ringing phones as soon as possible. The best way to accomplish this is with a loud alerting device that is difficult to ignore. A loudly ringing bell is a logical choice.

Now the question is, how to electrically ring the bell on cue? Actually this is not difficult to do by electromechanical means. The system that became the norm in virtually all telephones until suitable electronic circuitry was developed was invented by Thomas A. Watson—Alexander Graham Bell's assistant. He got the patent for this device in 1878.

The basic mechanism of the standard electromechanical telephone ringer is illustrated in Fig. 3-2. The simple circuitry

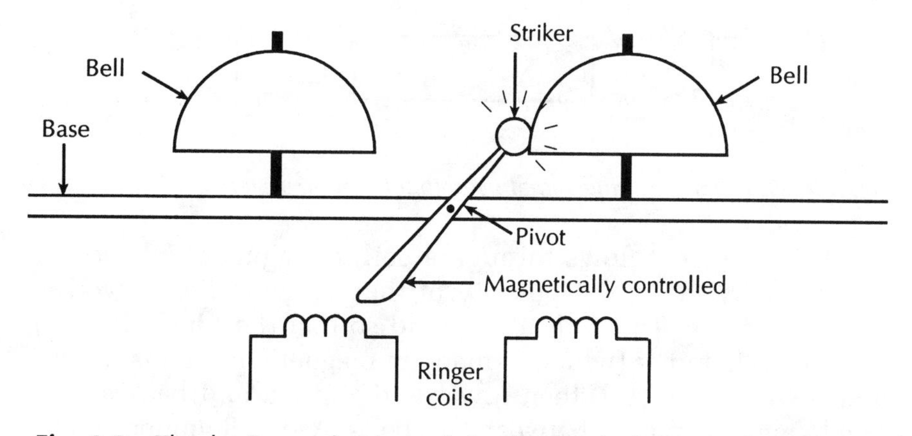

Fig. 3-2 *The basic mechanism of the standard electromechanical telephone ringer.*

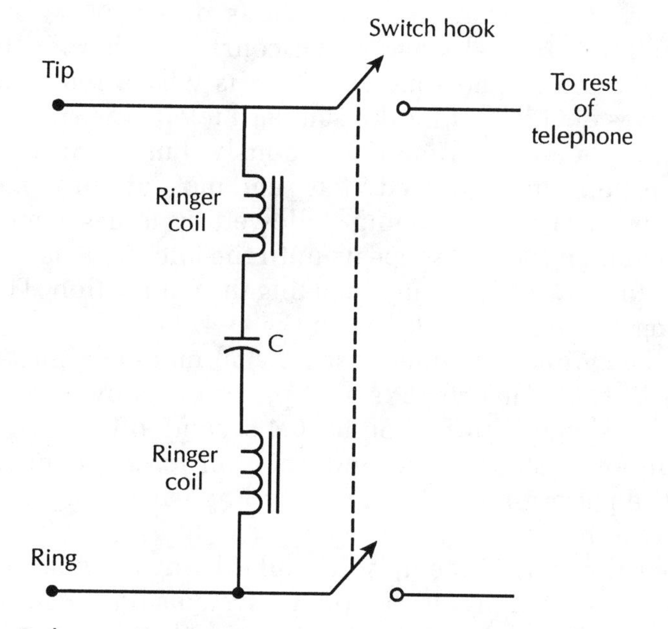

Fig. 3-3 *Only very simple circuitry is required to activate the standard electromechanical ringer.*

required to activate the electromechanical ringer is shown in Fig. 3-3. This is an incredibly simple circuit—just two coils and a capacitor. The capacitor blocks any dc from flowing through the ringer circuit. This prevents the ringer circuit from looking like a dead short from the ring to the tip wires, which would prevent the rest of the telephone from working when the switch hook is closed.

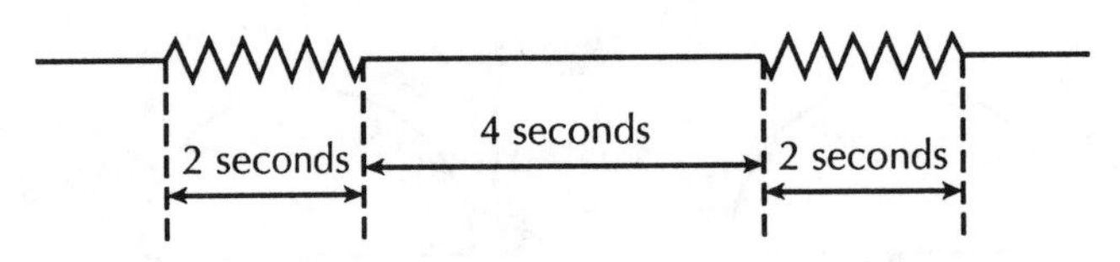

Fig. 3-4 *The standard ring signal used in telephone systems.*

When current flows through a coil, a magnetic field forms around it. With ac, the polarity of the magnetic field reverses twice per cycle. The coils are wound so that the pivoted armature of the hammer (which is made of magnetic metal) is drawn first towards one coil, then towards the other, and back again. The hammer is placed between two bells. As the hammer moves back and forth, it strikes the bells.

The higher the applied voltage, the larger the induced magnetic field. This means the hammer is drawn back and forth faster and hits the bells harder. Of course, the harder the bells are hit, the louder the sound is. This is why a relatively high voltage was decided on for the standard telephone ringer signal.

The bells are not rung continuously, but in an on and off pulsed pattern. In the United States and most of Europe, the pattern is two seconds of ringing, followed by four seconds of silence, and then the cycle repeats until the telephone is answered or the calling party hangs up, breaking the connection. This standard ringer signal is illustrated in Fig. 3-4.

In the United Kingdom, a somewhat more complex ringing pattern is used. This ringer signal pattern is shown in Fig. 3-5. The ringer signal is turned on for 0.4 second, off for 0.2 second, on again for another 0.4 second, then off for 2 seconds, before the entire pattern repeats. This gives English telephones a very distinctive ring.

Sometimes a different, nonstandard ringer signal might be used, especially in private telephone systems (PABXs). Different patterns can be used to indicate, for example, if the call is originating internally to the system (from another extension) or externally (an incoming call). The user can tell by the sound what type of call to expect when they answer their telephone.

Why does the telephone company bother with the extra circuitry required to pulse the ringer on and off? Wouldn't it be easier and more practical just to ring the customer's telephone continuously until it's answered or the call is discontinued? Easier, yes. But not more practical. The central switching office must

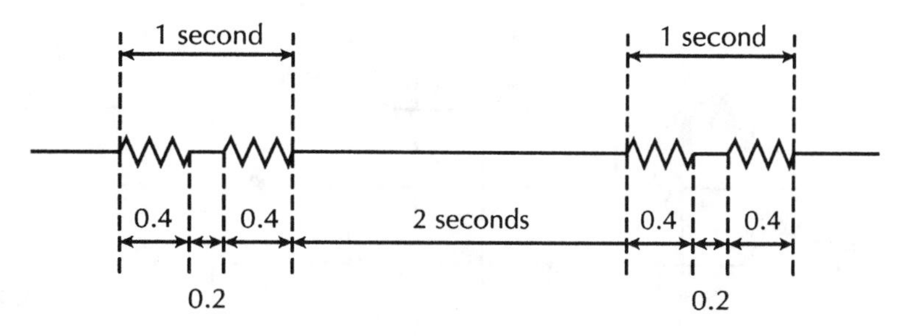

Fig. 3-5 *Telephone systems in the United Kingdom use a somewhat more complex ring signal pattern.*

handle a lot of calls simultaneously. Dozens or hundreds might be ringing at once. By sending the ringer signal continuously, an awful lot of power would be consumed. It is more efficient to pulse the signals in an overlapping pattern. Figure 3-6 shows how three ringer circuits could be used, each delayed by 2 seconds with respect to the others. At any given instant, only two of the generator circuits are actually sending out a signal. In an actual central switching office, there are more ringer circuits than just three, and the interleaving pattern is more complex, with shorter delays than in our simple illustration, but the principle is the same.

Another advantage of the multiple, interleaved ringer generators is that when a new incoming call is placed, the ringer signal can be sent out immediately. The system doesn't have to wait until the next on pulse period if the connection is made near the beginning of the off (pause) period. For any single call, the time savings of this are negligible, but cumulatively it can add up to quite a savings for the telephone company.

At the same time the central switching office is sending out the pulsed ringer signals, it is continuously monitoring the called telephone's switch hook. As soon as it senses the dc start to flow, indicating the switch hook has closed (the handset has been lifted from the cradle), the ringer signal turns off within 200 milliseconds (ms). This is known as the *ringtrip* process.

The telephone company's sensing of the off-hook condition is highly dependent on the impedance of the individual line circuit. Anyone designing add-on telephone equipment must be careful that their projects don't significantly throw off the overall system impedance, or their telephone service will be erratic, at best.

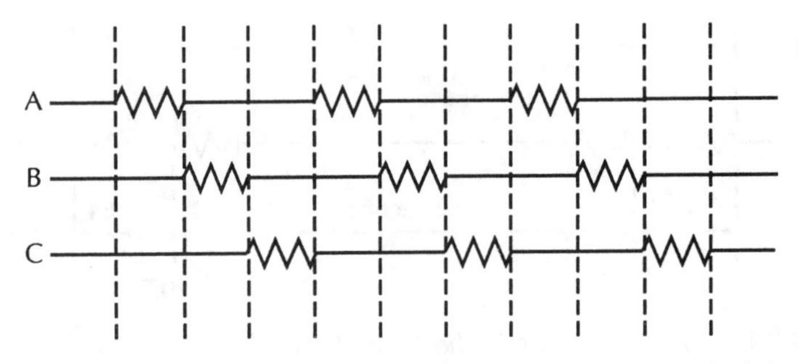

Fig. 3-6 Three ring circuits can be used simultaneously, each delayed by 2 seconds with respect to the others.

Rotary dialer

A typical rotary dialer circuit is shown in Fig. 3-7. The normally closed single-pole, double-throw (SPDT) switch (S1) is the switch hook. Usually, two push buttons are used, one on either side of the cradle. When the handset is hung up, both push-button switches are simultaneously activated. The weight of the handset holds the switch hook open, breaking the dc circuit between the rest of the telephone and the telephone lines.

The circuit can also be broken by opening normally closed switch S3. This switch is a part of the rotary dial mechanism. As the dial rotates backwards, fingers on the underside of the dial mechanically open and close this switch, once per digit. This regular interruption of the current flow can be detected and interpreted at the central switching office. The capacitor across the dial switch helps suppress voltage spikes from the rapid changes in the current flow as the switch is opened and closed with the rotation of the dial.

In modern telephones, the mechanical rotary dial is replaced by an electronic switching circuit, which serves the exact same purpose. Electrically both types of pulse dialer are identical. However, the electronic switching circuit can include a memory and other special features that are not possible with a mechanical rotary dial.

Touch-Tone® dialing is always accomplished via a purely electronic circuit.

Antitinkle and speech muting circuit

As the rotary dial rotates, the circuit is mechanically opened and closed. This can cause strong current surges that can feed back

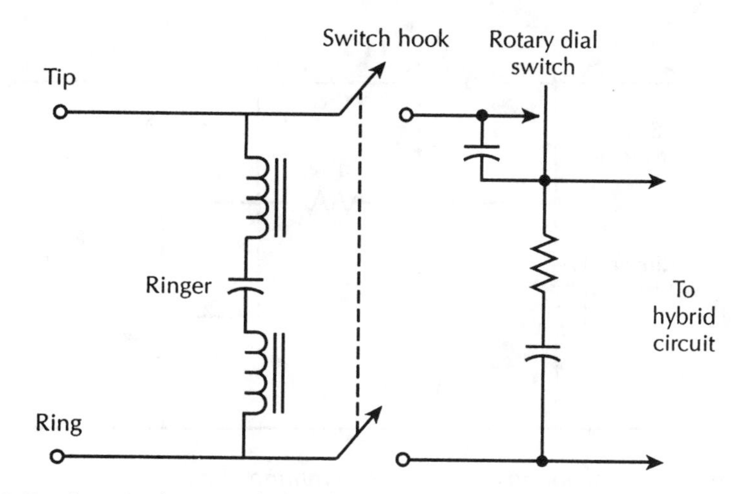

Fig. 3-7 *A typical rotary dial circuit.*

through the ringer circuit. The ringer circuit can't tell these switching pulses from a true ring signal, so the hammer is moved in one direction or the other on each spike pulse, banging one of the bells and making it ring. These stray rings are relatively soft, making a tinkling sound each time the dial is rotated. This isn't harmful, but it can be annoying. An antitinkle circuit is used to suppress the switching spikes and eliminate or significantly reduce the annoying sounding.

The antitinkle circuit is often combined with a speech muting circuit. With the speech muting circuit, the switching pulse spikes flow through the audio circuits and are loudly re-created in the earpiece. This can be even more annoying than the tinkling bell. In some cases, these nuisance clicks can even damage the speaker or part of the audio circuitry. Therefore, it is particularly important to suppress these clicks by muting the speech circuitry during the dialing process.

A fairly typical antitinkle and speech muting circuit is shown in Fig. 3-8. Notice that a single circuit simultaneously serves both purposes.

When the dial is rotated away from its normal, rest position, normally open double-pole, single-throw (DPST) switch S2 is closed, activating the antitinkle/speech muting circuit. The capacitor in series with the ringer coil(s) acts as a spark suppressor, reducing the tendency for the bell to tinkle. At the same time, the ringer coil is shunted with a resistor. The value of this resistor is usually about 340 Ω. Switch S2 also places a short across the audio speech circuit, effectively turning off this part of the tele-

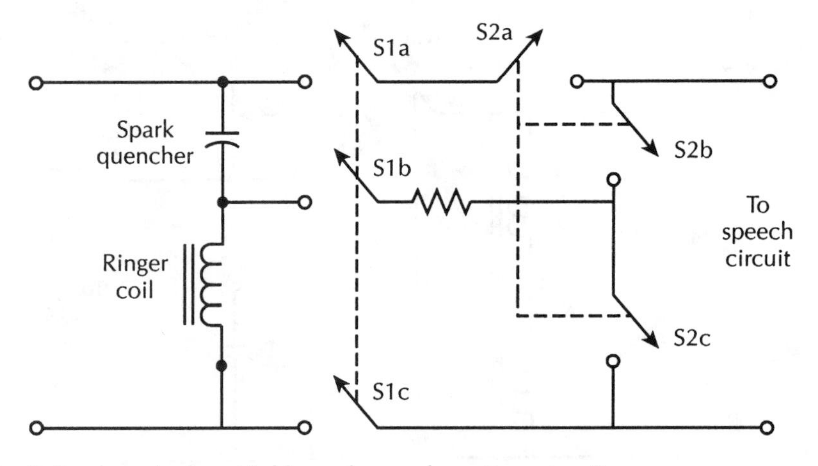

Fig. 3-8 *A typical antitinkle and speech muting circuit.*

phone so no sound is heard through the earpiece until the dial rotates back to its normal full-rest position, opening shunt switch S2.

The antitinkle and speech muting circuits are rarely needed when electronic pulse dialing circuits are used. They would serve no purpose in a Touch-Tone® telephone.

The handset

The telephone handset is a very simple device, even in modern, multifunction, all-electronic telephones. This part of the telephone has two important functions. It converts the audio of the user's voice into electrical signals to be transmitted over the telephone lines, and it converts received electrical signals back into audio signals (speech) that can be heard. Of course, anyone familiar with electronics knows that the first task can be accomplished with a microphone, while a speaker will do the second part of the job.

The handset consists of a microphone in the mouthpiece and a speaker in the earpiece, as shown in Fig. 3-9. That is all there is to it, other than the physical dimensions of the handset unit itself. It is designed so that when an average adult holds the earpiece by their ear, the mouthpiece is close to their mouth.

Because of the way the audio signals are used in a telephone system, the microphone is usually called the *transmitter*—this is where the speech signals to be transmitted enter the system. The speaker is called the *receiver*—the signals received through the telephone lines are heard here. In casual usage, the entire

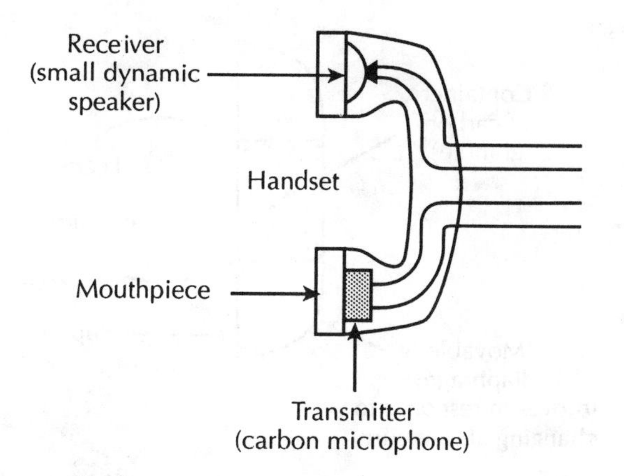

Fig. 3-9 *Most telephone handsets use a carbon microphone as the transmitter.*

handset is often called the "receiver." This is rather inaccurate and potentially confusing, but this terminology is in widespread popular use.

The type of microphone used almost exclusively for the telephone transmitter is the carbon microphone, illustrated in Fig. 3-10. The advantages of this type of microphone are that they are inexpensive and very durable. The carbon microphone also puts out a fairly strong electrical signal, well suited for use in a telephone system.

In most audio applications, the chief limitation of the carbon microphone is its rather restricted frequency range. But because telephone systems intentionally limit the frequency response anyway, as discussed earlier, this is not a problem in this particular application.

As shown in Fig. 3-10, the carbon microphone's main component is a small, two-piece capsule filled with a great many tiny granules of carbon. Carbon is a fair electrical conductor with relatively high resistance. Standard resistors are composed of carbon.

The front and back of the capsule are metallic plates that are electrically insulated from each other, except for the relatively high resistance path between them through the enclosed carbon particles. One end of the capsule is held fixed in place in the housing of the handset unit. The opposite end of the capsule is connected to a movable diaphragm. When you speak into the mouthpiece, the sound of your voice causes the diaphragm to vibrate. When it moves back, towards the center of the capsule, the enclosed carbon granules are forced closer together. When

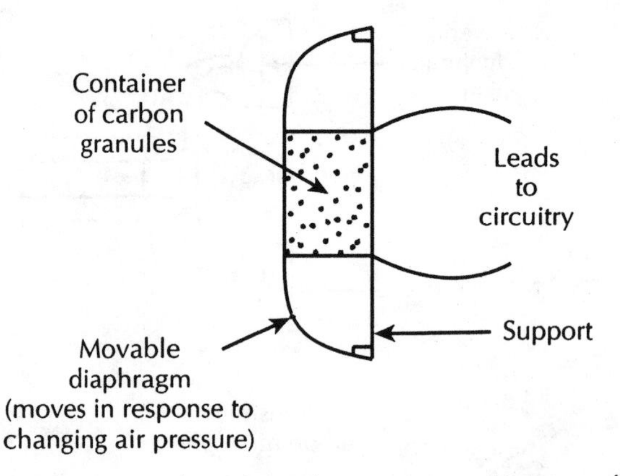

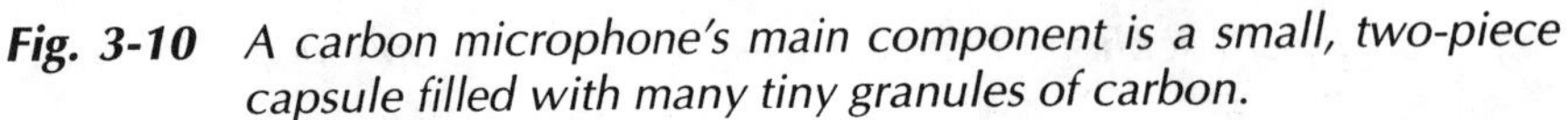

Fig. 3-10 *A carbon microphone's main component is a small, two-piece capsule filled with many tiny granules of carbon.*

the diaphragm moves in the opposite direction (away from the center of the capsule), the granules move apart. The relative density of the carbon particles determines the overall resistance between the capsules front and rear plates, so the resistance varies with the vibrations of the diaphragm. The capsule is electrically connected to a dc voltage source—in this case, the supply voltage provided by the telephone company. This voltage is constant.

Ohm's law comes into play here:

$$I = E/R$$

Because the voltage (E) is constant, and resistance (R) varies with the sound pressure (vibrations of the diaphragm), the current (I) must be a direct electrical analog of the audio signal.

The speaker (receiver) located in the earpiece of the handset works in just the opposite manner. It converts the electrical signals received from the telephone lines into acoustic energy; that is, audible sounds. In this case, the sounds you hear are reproductions of the speech of the party on the other end of the line.

A standard speaker is an electromagnetic device. It consists of a movable diaphragm, a permanent magnet, and a coil (acting as an electromagnet). The coil, called an *armature* in this application, is attached to the center of the flexible diaphragm, as shown in Fig. 3-11. The permanent magnet's position is fixed (within the handset), and it provides a constant bias field for the varying electromagnetic field surrounding the coil to work against. This varying electromagnetic field is in direct propor-

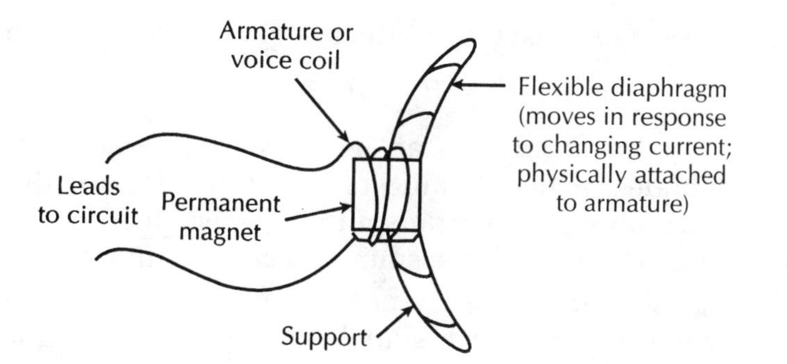

Fig. 3-11 *An armature coil is attached to the center of the flexible diaphragm in a speaker.*

tion to the electrical signal fed through the coil. The polarity of the varying electromagnetic field is constantly changing back and forth because an audio frequency signal is passing through the coil. When the polarity of the variable electromagnetic field is the same as that of the permanent magnet, the coil armature is forced to move away, pushing the attached diaphragm outwards. Similarly when the polarity of the variable electromagnetic field is in opposition to that of the permanent magnet, the coil armature is pulled towards the permanent magnet, pulling the attached diaphragm inwards. Consequently the diaphragm moves back and forth in step with the received electrical signals, reproducing the voice of the person on the other end of the telephone line.

Loop length compensation

Most of the sections in a telephone are pretty obvious in their function, but the loop length compensation circuit probably requires a little explanation. As in any electrical circuit, resistance and impedance (ac resistance) are critical factors. Unfortunately in a telephone system, they aren't entirely predictable or consistent. There are three major resistance components you need to be concerned with:

- R_c—resistance of the telephone's internal circuitry;
- R_t—varying resistance of the transmitter (or microphone); and
- R_l—resistance of the telephone lines between the central switching office and the telephone.

The total effective resistance of the system can be expressed as

$$R_x = R_c + R_t + R_l$$

Of these three, only R_c can be considered a constant. For most telephones, it has a value of about 400 Ω. For the time being I will ignore R_l. The transmitter's resistance is intentionally designed to vary with the sound level at the mouthpiece. By Ohm's law this varying resistance is converted into a varying current, which is the speech signal carried from one telephone to the other:

$$I = E/R_x$$

In this case, E is the voltage supplied on the telephone lines. This voltage is nominally 48 V. The actual value will vary quite a bit. This isn't significant for our discussion. The telephone company's central office directly monitors its operating voltage and compensates accordingly. For a given telephone conversation, the voltage can be considered a constant value, and can reasonably be assumed to be the nominal value of 48 V:

$$\begin{aligned} I &= 48/R_x \\ &= 48/(R_c + R_t) \\ &= 48/(400 + R_t) \end{aligned}$$

Because R_t varies with the speech signal, the current will be proportionate to the speech signal at all times. The telephone company (and the telephone at the other end of the line) can decode the electrical signal and re-create the original speech signal. It's just a matter of solving for one unknown variable, because all other elements in the equation are known, once I is measured by the telephone equipment. Mathematically the equation can be rearranged to

$$R_t = (48/I) - 400$$

This is simple and straightforward enough. But things get a bit more complicated when R_l (the resistance of the cables connecting the telephone to the central switching office) is added to the equation:

$$I = 48/(R_c + R_t + R_l)$$

The value of R_l is different for every telephone in the system. Cable has a specific resistance per foot. The longer the wire, the higher the resistance. This means the further a given telephone is from the central office, the higher its R_l value must be. If this is not compensated for, there will be an awful problem balancing the volume levels for all telephones in the system.

Let's assume there are only three telephones of interest in the system—call them A, B, and C. Phones A and B are 5 miles away from each other. Phone C is another 5 miles in the same direction from B, so C is 10 miles away from A.

Let's say A calls B. No problem. B calls C. Again no problem. Because the line distance is the same for each call, the volume of the transmitted speech is the same in each case (assuming the callers speak at more or less the same levels). But what happens when A calls C? The distance is twice as far, so R_l must be twice as much. In the first two calls, the current was equal to:

$$I = 48/(R_c + R_t + R_l)$$

Doubling the value of R_l is the equivalent of rewriting our equation as

$$I = 48/(R_c + R_t + (2 \times R_l))$$

Obviously the current this time must be significantly lower than the earlier calls. This means either C will hear A's voice at too low a volume or A will have to speak louder.

To eliminate such problems, a loop length compensation is included in modern telephones. A simplified circuit diagram is shown in Fig. 3-12. The component you are most interested in here is the varistor. For a shorter loop length (less wire between the telephone and the central office), the current is increased (because the line resistance is decreased). The varistor senses this and proportionately lowers its resistance in response. This shunts some of the line current around the transmitter until the transmitter current is approximately the same as the line current. The varistor is now balanced. Electrically the telephone "looks" like it is on a longer loop (further from the central office.)

Now let's move that same telephone to a new installation, further from the central office. The line resistance (R_l) is increased by the longer wires, decreasing the line current. The varistor increases its resistance, applying more of the line current to pass through the transmitter. The effective transmitter current is boosted, so it is again balanced with the line current.

The compensation provided by the varistor minimizes the effect of the distance of the telephone from the central office. This results in a more consistent signal volume, regardless of how far apart the two telephones in the connection are from one another.

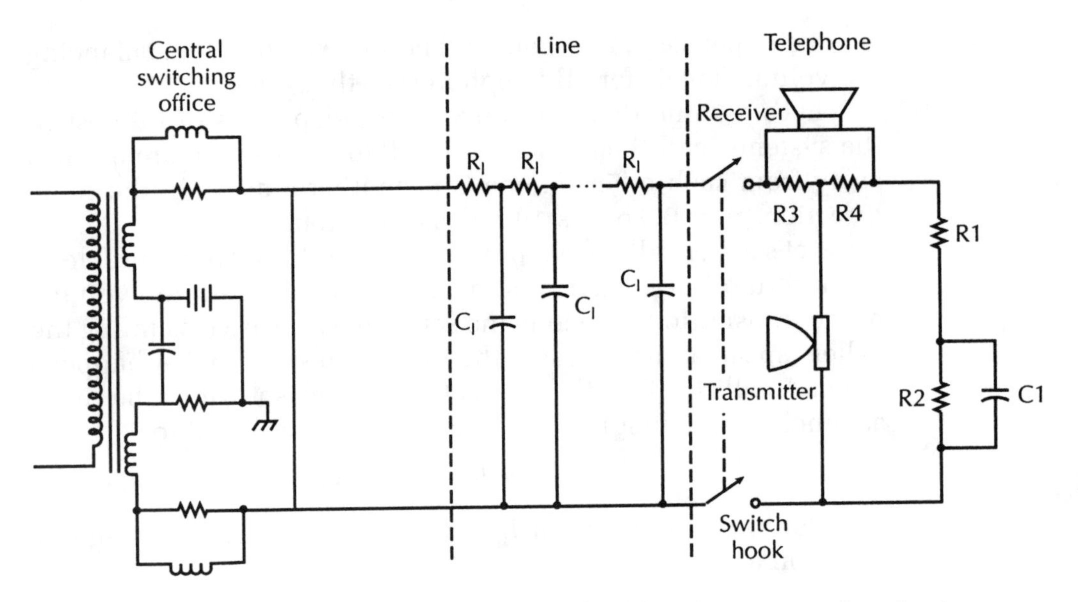

Fig. 3-12 *Most modern telephones use a loop length compensation circuit.*

The hybrid circuit

The hybrid circuit is often called the *induction coil,* a name that was more appropriate in early telephones, but the name has carried over into modern technical usage.

You should recall that the telephone uses four connection wires (red, green, yellow, and black), but there are only two wires between the telephone company's central office and the user's telephone. Now, how can you get four connections out of two wires, or vice versa? This is the function of the hybrid circuit.

Many communications systems, including modern telephones, use a system called *full duplex.* This means signals can be sent in both directions over the same wire at the same time. Two-wire cables to connect subscriber's telephones into the system are a lot less expensive than four-wire cables, and the hybrid circuit permits the use of the cheaper connecting cables.

Basically the hybrid circuit is a multiwinding transformer, as illustrated in Fig. 3-13. In most practical telephones, a single all-in-one transformer is not used, because it would be expensive to manufacture properly. Instead, hybrids normally consists of a pair of interconnected transformers contained within a single housing.

The various sections of the hybrid function much like any

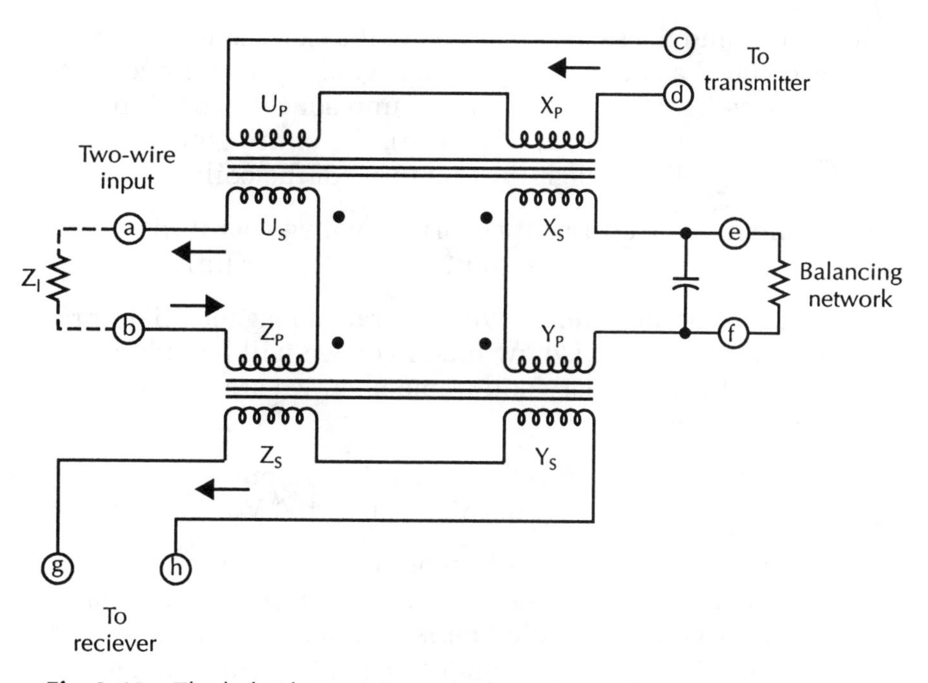

Fig. 3-13 *The hybrid circuit in a telephone is usually a multiwinding transformer.*

standard transformer. A transformer consists of two (or more) separate coils, usually wound around a common core. The windings of the two coils are placed very close to each other. In most cases they are intertwined. But the windings of the two coils are electrically insulated from one another. They usually aren't used in the same circuit loop. The two windings in a transformer are called the *primary* and the *secondary*. For all intents and purposes, the primary is the transformer's input and the secondary is its output.

When an ac voltage is fed through the primary, a fluctuating magnetic field is created around the windings. As this magnetic field moves back and forth (with the changing polarity of the ac voltage), it cuts across the windings of the secondary. This induces a voltage in the secondary coil that is proportionate to the original voltage in the primary coil, even though there is no direct electrical connection between the two coils in the transformer.

The ratio of the input and output voltages is determined by the ratio of the number of turns in the primary and secondary windings. For example, let's imagine a typical transformer with 200 turns in its primary and 200 turns in its secondary. The primary coil and the secondary coil are the exact same size. Now let's feed 100-V ac across the primary winding. Assuming there are no losses in the transformer (an unreachable condition for any real-world transformer, of course), the entire 100 V will be distributed equally through the 200 turns of the coil:

$$\text{input voltage/turns (primary)} = \text{voltage per turn}$$
$$100/200 = 0.5 \text{ V per turn}$$

Each turn of the primary will generate a magnetic field proportional to 0.5 V. This is how much voltage will be induced in each turn of the secondary coil (remember, we are assuming no losses in the system):

$$\text{volts per turn} \times \text{turns (secondary)} = \text{output voltage}$$
$$0.5 \text{ V} \times 200 = 100 \text{ V}$$

In this case, the output voltage is equal to the input voltage, because the primary and secondary windings have the same number of turns. This is called an *isolation transformer* or a 1:1 transformer. It is used for impedance matching and/or to electrically insulate two circuits from one another.

Most actual transformers have different numbers of windings in the primary and the secondary. For example, let's again assume there are 200 turns in the primary winding, but this time there are only 50 turns in the secondary winding. Again, assume the input voltage is 100-V ac. The primary coil will function in exactly the same way as before:

$$\text{input voltage/turns (primary)} = \text{voltage per turn}$$
$$100 \text{ V}/200 = 0.5 \text{ V per turn}$$

But the secondary winding has fewer turns, so a smaller voltage is induced into it because the voltage per turn remains the same:

$$\text{volts per turn} \times \text{turns (secondary)} = \text{output voltage}$$
$$0.5 \text{ V} \times 50 = 25 \text{ V}$$

This is called a *step-down transformer* because the input voltage (100 V) is stepped down to a lower output voltage (25 V).

A third possible type of transformer is the *step-up transformer,* which has more turns in the secondary winding than in the primary winding. This time, assume the primary winding consists of 200 turns and the secondary winding consists of 300 turns. As in the earlier examples, assume the input voltage is 100-V ac:

$$\text{input voltage/turns (primary)} = \text{voltage per turn}$$
$$100\ \text{V}/200 = 0.5\ \text{V per turn}$$

$$\text{volts per turn} \times \text{turns (secondary)} = \text{output voltage}$$
$$0.5\ \text{V} \times 300 = 150\ \text{V}$$

Notice that this time the output voltage is higher than the input voltage. No, you aren't getting something for nothing here. The power in the primary and secondary windings must remain constant in each case. Increasing the output voltage decreases the output current, and vice versa. The power (voltage × current) is the same in all cases. Comparison of the three types of transformers is shown in Table 3-1.

For simplicity in this discussion, the possibility of losses in the transformer has been ignored. In any practical transformer, some of the input power is consumed by the transformer action, decreasing the output power somewhat. To get the output voltages in the examples, you would probably need to add a few extra turns to the secondary windings. For example, for a step-down transformer to drop 100 V to 25 V, with 200 turns in the primary winding, it might actually need 54 or 56 turns, instead of the nice, even 50 turns used in our ideal example.

A transformer can be used backwards, reversing which coil acts as the primary and which functions as the secondary. The only difference between a step-down transformer and a step-up transformer lies in which winding (the larger or the smaller) is used as the primary with the input voltage applied across it.

The hybrid circuit is essentially several simple transformers used together. One of its primary functions is impedance matching to reduce power loss and signal reflections in the telephone system. Let's look at the hybrid circuit of Fig. 3-13 in a little more detail.

Notice that there are four transformer sections, labeled U, X, Y, and Z. (The letter V is skipped to avoid possible confusion with voltage.) For convenience, the winding that acts like the

Table 3-1 Comparison of basic transformers.

Type of transformer	Output voltage compared to input voltage	Output current compared to input current
1:1	Same	Same
Step-down	Lower	Higher
Step-up	Higher	Lower

primary in this circuit is marked with a subscript p. Similarly, a subscript s indicates a functional secondary winding. The dots are polarity indicators. If one transformer section has its polarity mismatched, it will work against, rather than with the rest of the circuit.

The input signal from the telephone lines is fed into Z_p, and is induced into Z_s, and is then fed into the receiver (the speaker in the earpiece of the handset). The signal from the transmitter (microphone in the mouthpiece of the handset) is fed into U_p, where it is induced into U_s, and fed out through the telephone lines. The transmitter's signal is also fed across X_p, so it is induced into X_s, which feeds the signal into an impedance-matching balancing network. Then the balanced signal is applied across Y_p to be induced into Y_s in the receiver circuit. This echoes some of the transmitter signal (the user's voice) back into the user's own earpiece.

It feels very unnatural to speak on the telephone without this acoustic feedback. Studies have indicated that most people speak too loudly without the echo. The level of the echo must be carefully set by the balancing network. If it is too loud, the speaker will tend to use a lower voice. The effect of echo feedback is usually subliminal, but it does make a definite difference in the way people use telephones. This echo signal is usually called the *sidetone* by telephone technicians.

A similar hybrid circuit is also used at the opposite end of the line in the telephone company's central switching office. The only difference is that this hybrid circuit's balancing network is adjusted so that there is no sidetone. None of the transmitter signal is fed back into the receiver circuit. The sidetone is strictly a function of your own telephone. The hybrid circuit in the central switching office is also physically larger so it is able to handle the heftier current levels it is subject to.

Modern electronic telephones

Early telephones relied mainly on electromechanical components. Few, if any, active components (those capable of amplification) were used in telephone devices until recently. Today, old-style electromechanical telephones are a rarity. Almost all modern telephones use active electronic circuitry. The greater versatility of electronics permits many new functions, but everything must be designed to be fully compatible with the older electromechanical systems.

There would be little point in attempting to build your own telephone from scratch. It would be much more expensive than buying a commercial unit, and a hobbyist probably couldn't achieve the small, compact circuitry of commercial, assembly line manufacturers. Your home-brew project would almost inevitably be rather clunky by modern standards, unless you are a real expert. And if you're that good at hand-building electronic projects, why not use your talents for more interesting and worthwhile projects than to duplicate inexpensive, widely available units?

The following chapters include a number of telephone-related projects that I think might be worth your while. They either do things that commercial equipment doesn't do, or they duplicate functions of commercial equipment that is rather expensive. The big advantage to being an electronics hobbyist is you can customize your system to suit your own individual needs.

❖4 Ringer projects

Every telephone has a built-in ringer of some sort. So why would anyone want to bother building their own ringer projects? Actually there are a number of perfectly good reasons to do so.

Perhaps the most obvious reason is to replace a defective ringer in an otherwise good telephone receiver. A more common reason is to replace an existing ringer because you don't like the way it sounds. The old-fashioned electromechanical bell included in older telephones can be very jarring and disturbing. An electronically generated tone can alert you to an incoming telephone call in a much more pleasant manner.

You can also "personalize" the sound of your telephone's ringer. This can be done just for fun, or it can serve a very practical purpose in an environment with multiple telephone lines, such as in a large office. If all the lines ring with the same sound, how can you tell which telephone is ringing? By the sound alone, you can't tell. Usually multiline telephones have blinking lights to indicate which line (or lines) is waiting to be answered, but this isn't much help if you can't see the telephone when it starts to ring. A distinctive ringer sound for each extension or line in the system eliminates the need for interruptions and inconveniences. You can tell by the sound of the ring (which might not sound at all like a ringing bell) whether an incoming call is for you. If it's not your distinctive ring signal, you can ignore it and carry on with your work without interruption.

Another reason for wanting to build a telephone ringer project is to hear when the telephone rings and you are some distance away. For example, if you live in a large house, you probably don't need or want a telephone in every room. But in some parts of the house, it might be difficult to hear the telephone when it rings. You can install an extension ringer. It is less expensive and smaller than a complete telephone extension, which isn't really needed except for its ringer. An extension ringer circuit is hooked up to the telephone lines in exactly the same way as an ordinary extension telephone.

An extension ringer can be even more helpful outside or in outbuildings (such as a garage or a toolshed). Any connecting cable used outside must be heavily insulated and weatherproofed. Use only cable rated for outdoor use. An outdoor extension ringer circuit (actually any electronic project used outdoors) must be enclosed in a waterproof, weather-resistant housing. Try to mount the unit where it will not be subjected to adverse environmental conditions. For example, mount it under an extending eave.

As you can see, there are a number of perfectly good reasons for wanting an extension ringer circuit. I'm sure there are many special applications I haven't mentioned here. Whatever your individual need might be, this chapter presents several practical ringer projects for you to choose from.

The Federal Communications Commission (FCC) and the Electronics Industries Association (EIA) have issued several operating specifications that must be met by all ringer circuits in order for them to be legally connected to the telephone lines. These requirements are in no way unreasonable or overly restrictive. They just ensure that the telephone system will function properly and that service won't be disrupted by a nonstandard device connected to the telephone lines. The most important specifications concern the input impedance of the ringer device, which affects the overall loading on the telephone lines.

The FCC and EIA requirements also state that a ringer circuit must function when a ring signal is present, yet it should not sound when no ring signal is present. That is, it should not be falsely triggered by dialing signals, voice signals, or ordinary, predictable noise signals. (No one can completely guarantee against an unusual and unpredictable fluke noise spike.) In other words, these requirements demand that the ringer circuit must perform its intended task with reasonable reliability. Actually such official requirements should be unnecessary. Why on earth would anyone want to bother with a ringer circuit that doesn't operate reliably?

Powering a ringer project

Unlike most electronic projects, telephone projects often require no external power supply. This is because the telephone lines themselves carry a usable voltage that can be tapped. This is one reason why the telephone company wants to know what you've got hooked up to their lines. Because cumulative loading and other problems can affect telephone service, they have a legitimate reason to ask for this information. Refer back to chapter 2 for more information on the legal aspects of connecting anything to the telephone lines.

Remote or alternate ringers are almost always fully parasitic, taking their operating power entirely from the telephone line voltages. This is a logical choice, because the ringer circuit only needs to be activated when a higher-than-normal ring signal is sent along the phone lines. The rest of the time, the ringer circuit just sits there, not doing anything.

The connection for a ringer circuit is made between the tip and ring wires in the telephone cable. The ring wire is not dedicated to carrying signals that cause the telephone to ring. It refers back to the ring and tip on the old-style phone plugs used in early telephone switchboards. The terminology is obsolete, but the tip and ring designations are in too widespread use to disappear or change anytime in the foreseeable future.

The tip wire should be red and the ring wire should be green. These are standard color codes that should be followed in all telephone work. Watch out, though. Every now and then you might come across a cable that was put in by some careless technician who ignored the standard color code.

Only the ring and tip wires are relevant for a ringer project. Standard telephone cables have two additional wires that can just be ignored. No connection is made between the ringer circuit and either of these wires.

The tricky part in powering a ringer circuit (or other telephone project) is that a fairly large dc voltage is always present between the tip and ring connections, even when the telephone is on the hook. In most modern telephone systems, this on-hook voltage has a nominal value of 50-V dc. When the telephone company wants to alert you that you have an incoming call, it sends a somewhat higher ac voltage along the tip and ring wires. In most modern telephone systems, the ring voltage is somewhere between 85 and 125 V peak-to-peak. This signal is illustrated in Fig. 4-1.

Fig. 4-1 *In most modern telephone systems, the ring signal is an ac voltage between 85 V and 125 V peak-to-peak.*

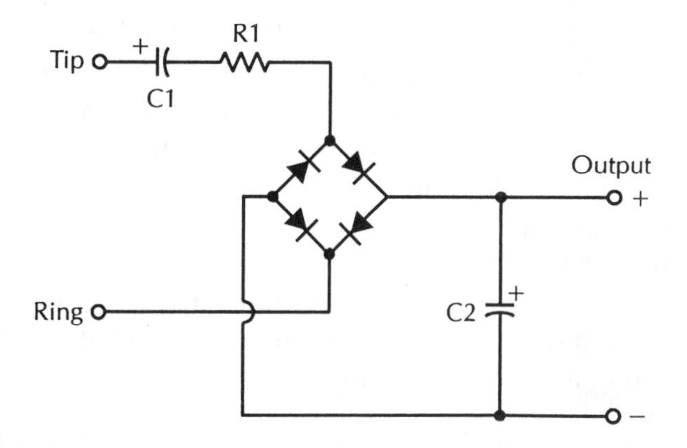

Fig. 4-2 *A simple, standardized ringer interface circuit.*

The ringer circuit must ignore the normal on-hook voltage and respond only to the ring signal voltage. Most practical ringer devices require a dc (or near dc) supply voltage. Fortunately, it isn't too difficult to find a solution. Fig. 4-2 shows a simple, standardized ringer interface circuit. This circuit or simple variations on it are used in each of the ringer projects presented in this chapter.

Because you are dealing with a fairly hefty ac voltage (85- to 125-V ac), adequate shielding is a must. Be very careful in the layout of your circuit. Try to prevent any possibility of short circuits forming by a wire touching some conductive surface connected to another part of the circuit. The interface circuit must be completely enclosed and fully insulated. It should be impossible for anyone to accidentally touch a live wire while the project is in use. Please don't cut corners when it comes to safety. It's never worth the risk.

This interface circuit is not particularly complicated. The input capacitor (C1) blocks the on-hook dc voltage and passes only the ac ring signal. The resistor serves as a simple current limiter. It can be omitted in some ringer circuits, although in most cases this isn't a very good idea. The telephone company will hold you responsible if a short or other problem in your project draws excessive current and causes loading problems in the telephone lines. You don't want to risk affecting the telephone service of your neighbors.

The ac ring signal is rectified by a bridge rectifier network made up of four ordinary silicon diodes. Almost any standard silicon diodes will work in this application. The specifications are not critical. To avoid premature diode burnout, however, the PIV rating for each diode in the bridge should be at least 200 V. A higher PIV rating would serve as extra insurance, but is not essential. Of course, a dedicated bridge rectifier unit (with a PIV rating of at least 200 V) could be used in place of four separate diodes, if you prefer.

The output capacitor (C2) filters some of the ripple out of the rectified ring signal. The ideal value for this filter capacitor will depend somewhat on the specific ringer circuit being used. Typically this capacitor will have a value of about 10 to 20 microfarads (μF).

When the ring signal appears on your telephone line, the ringer circuit receives power and is activated, but at other times it is deactivated. You should be aware that sometimes noise signals can cause brief false ring signals. In systems using old rotary-style phones, the quick making and breaking of the connection as the dial rotates can cause some ac signal to get through the filter capacitor, causing the ringer to emit short pulses of sound in step with the dial pulses. This is rarely a significant problem, but it can be rather annoying. Some more sophisticated ringer circuits might have provisions to prevent such problems. Relevant ideas on reducing such annoyances are offered in the text describing the individual projects in this chapter.

It is vitally important to remember that whenever you are working with any ringer circuit or other telephone project, you are dealing with fairly hefty ac voltages throughout the circuit. The ac ringer signal on the telephone lines is every bit as much an electrical shock hazard as your household ac sockets. Notice that the nominal voltages in each case are not all that dissimilar—the telephone ringer voltage is about 90-V ac, while household ac power is 110 to 120 V. Treat the telephone ringer signals with the same respect and care you would use with an ordinary ac power supply. As far as the physical properties involved are concerned, there isn't any practical difference between the two.

Always watch out for possible shock hazards. Do not touch any bare, exposed wires in any electronic circuit connected to the telephone lines—you never know when a ringer signal or a large noise spike might come along. Be sure to completely enclose your finished project in a plastic or other nonconductive case. It should not be possible (whether by accident or by deliberate intent) for anyone to touch any potentially live wires or connection points in the circuit while it is in operation.

Simple remote ringer

A very simple remote ringer circuit is shown in Fig. 4-3. I doubt that a simpler ringer circuit than this could be designed. A suitable parts list for this project appears as Table 4-1.

The sounding device in this project is a simple solid-state buzzer of the type often used in low-volume alarm circuits and science fair projects. Radio Shack usually carries suitable buzzer units for just a couple of dollars.

If you prefer, you can substitute some other sound-generating device in place of the simple buzzer used here. Watch out for the correct supply voltages and polarities.

When a dc voltage is applied to a buzzer of this type, it pro-

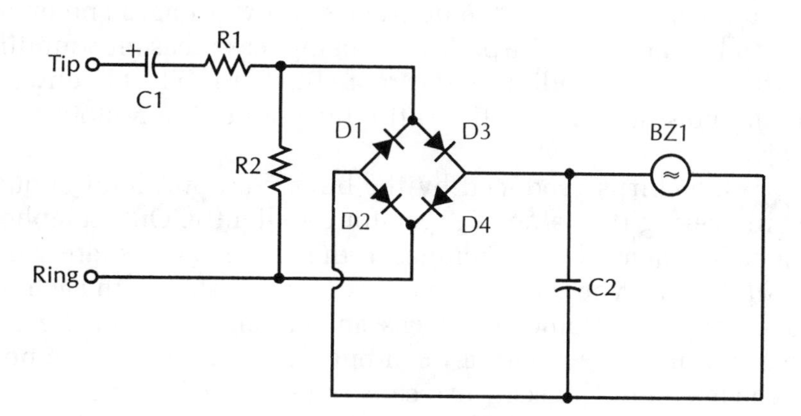

Fig. 4-3 *The schematic for a very simple remote ringer project.*

Table 4-1 Parts list for the simple ringer project of Fig. 4-3.

D1–D4	Diode (1N4003 or similar)
BZ1	Piezoelectric buzzer
C1	0.1-μF capacitor (see text)
C2	10-μF, 35-V electrolytic capacitor
R1	1.5-kΩ, 1/2-W, 5% resistor
R2	1-kΩ, 1/2-W, 5% resistor

duces a continuous tone. Most common buzzers are designed for use with voltages in the 2- to 12-V range. The continuous buzzing tone can be rather obnoxious, but if the buzzer is intermittently sounded in an on-off pattern, a much more pleasant chirping effect can be achieved. Fortunately, in this application, you don't have to do anything to get this chirping effect. It comes automatically, courtesy of the telephone company.

When there is an incoming call for your number, the telephone company sends a ring signal along the telephone line to activate the ringer device in the phone. This is usually an ac voltage of about 60 to 80 V, with a frequency of 20 Hz. When this ring signal is rectified by diodes D1 through D4, the result is a pulsating dc signal, which is filtered by capacitor C2. The pulsations in the buzzer's supply voltage cause it to chirp while the ring signal is present on the telephone lines. When the ring signal is absent, the buzzer remains silent.

You might want to experiment with other values for the filter capacitor (C2). Do not make this capacitor smaller than about 10 μF, however. If this capacitor's value is too low, there won't be sufficient filtering of the buzzer's supply voltage. The buzzer will tick rather than chirp. Besides being less pleasant sounding, the ticking is less audible and is more likely to be lost in environmental noises, defeating the entire purpose of the remote ringer project.

If the chirps produced by the buzzer are not loud enough, try increasing the value of capacitor C1 slightly. On a telephone line using rotary dial telephones, if the value of C1 is larger than about 0.25 μF, you will hear a quick chirp in step with each dial pulsation when someone places an outgoing call on the line. This will not occur with a push-button telephone. Static noise on the telephone line can also cause stray chirps from the buzzer if the value of capacitor C1 is too large. There is a trade-off between the chirp volume and excessive sensitivity to fluctuations on the telephone line.

Never use a capacitor larger than 0.5 μF for C1. The buzzer and possibly even the telephone line itself could be damaged. If you need a louder signal, replace the buzzer with an oscillator circuit and amplify the output signal as needed. For most practical purposes, however, a standard buzzer is plenty loud enough. In fact, you might find it annoyingly loud.

A mylar capacitor or similar high-grade device is strongly recommended for C1. It should be rated for a working voltage of at least 100 V, but a capacitor with a higher working voltage will be much less prone to failure. Remember, noise fluctuations and excessive voltage spikes can occur on the telephone lines. The full incoming voltage is passed through this capacitor. A 500-V capacitor would be a very good choice here.

The value for resistor R1 is also rather critical in this circuit. Do not change this resistor's value more than 10 percent in either direction from the 1.5-kΩ (1500 Ω) resistor recommended in the suggested parts list for this project. Definitely use a 1/2-watt (W) resistor here. A 1/4-W resistor runs too great a risk of being overloaded by the ring signal.

Almost any standard silicon diodes will work in this circuit's bridge rectifier (D1 through D4). There are no special requirements here. The diodes used should all be of the same type number. They should also have a PIV rating of at least 200 V. A higher PIV rating certainly wouldn't hurt. Diodes with a PIV rating lower than 200 V are likely to blow out because of noise voltages that appear on the telephone lines under some conditions.

The MC34012 telephone ringer IC

Telephone ringer applications are popular enough to warrant several dedicated integrated circuits (ICs) designed precisely for such applications. One of the most popular telephone ringer chips is the MC34012 from Motorola. This is a simple eight-pin dual in-line package (DIP) device, as illustrated in Fig. 4-4.

As the "MC" prefix in its identifying number indicates, the MC34012 IC is manufactured by Motorola. Similar chips are available from other semiconductor manufacturers, but I will

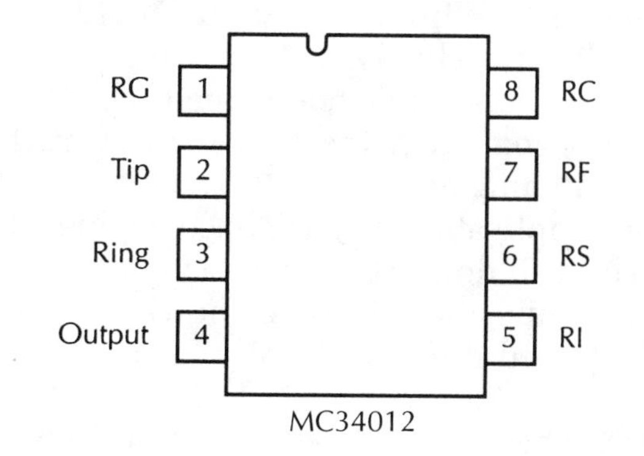

Fig. 4-4 *The MC34012 telephone ringer IC can be used in many useful projects.*

consider only the MC34012 here because it seems to be the most readily available of such units.

The MC34012 is available in several forms, identified by a suffix number. The MC34012-1 is designed to generate an output signal with a frequency of 1000 Hz. The MC34012-2's output frequency is 2000 Hz. You'd probably expect the MC34012-3 to have an even higher output frequency, but it doesn't. The output frequency for the MC34012-3 is 500 Hz, the lowest available output frequency. These are the nominal output frequencys for each version of the MC34012. The actual output frequency can be varied somewhat, if necessary, by external circuitry.

This IC is specifically designed to drive sounding devices such as common piezoelectric buzzers. A big advantage of using a chip like this in a ringer project is that its internal circuitry is designed to meet all of the operational requirements required by the FCC and EIA for a ringer circuit. This takes some of the fuss and bother out of designing a specialized ringer project. The circuit designer doesn't have to worry about such matters, because they are already taken care of by the IC's manufacturer.

The MC34012 does not require any external power source. It "steals" its operating signal from the telephone lines—specifically from the ringer signal itself. Most ICs are designed to operate only on dc supply voltages, but this one is intended for use with the ac ringer signal appearing on telephone lines. As you should remember from earlier chapters, the standard ringer voltage is about 90-V ac—a very high value for any IC to handle. Moreover, it can vary a great deal, depending on the area and the present load on the local telephone lines. The MC34012, besides being able to handle such a high voltage level, is quite forgiving when it comes to voltage fluctuations. It can cheerfully operate off of anything ranging from 24-V ac up to 120-V ac. In other words, it should work reliably with any ringer signal appearing on any standard, nondefective telephone line.

The output signal from the MC34012 is in the form of a square wave, switching between +10 and −10 V (20 V peak-to-peak). For some applications, this voltage might be too high, and you'll need to bring it down some with a voltage divider network of some kind. For most applications, however, the output from this chip can be used directly, usually to sound a piezoelectric buzzer.

Remember, when working with any project involving the MC34012, you are dealing with fairly hefty ac voltages throughout the circuit. Watch out for possible shock hazards. Do not touch bare, exposed wires when the circuit is connected to the telephone lines—you never know when a ring signal or a large noise spike might come along. A MC34012 probably isn't capable of giving a healthy person a fatal shock (although this certainly can't be guaranteed in any ac circuit), but it can be very painful and could cause serious injury. Please be careful and don't take any foolish chances.

Be sure to completely enclose your finished project in a plastic or other nonconductive case. It should not be possible for anyone to touch any potentially live wires or connection points in the circuit while it is in operation.

The risk of electrical shock is too serious to ignore or to take unnecessary chances. The ac ringer signal on the telephone lines is every bit as much an electrical shock hazard as your household ac sockets. Notice that the nominal voltages in each case are not all that dissimilar—the telephone ringer voltage is about 90-V ac, while household ac power is 110 to 120 V. Treat telephone ringer signals with the same respect and care you would use with an ordinary ac power supply. As far as the physical properties involved are concerned, there isn't any practical difference between the two.

The MC34012 telephone ringer IC is quite easy to use. The basic ringer circuit for the MC34012 is shown in Fig. 4-5. A suitable parts list for this simple but effective ringer project is given as Table 4-2. You are encouraged to experiment with alternate component values in this circuit.

As in most telephone projects, no external power supply is required in this circuit. The circuit's power is taken directly from

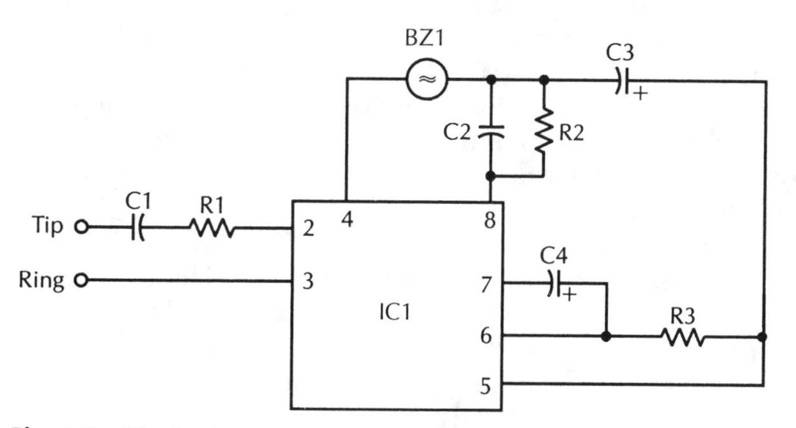

Fig. 4-5 *The basic ringer circuit for the MC34012.*

Table 4-2 Parts list for the MC34012 ringer project of Fig. 4-5.

IC1	MC34012 telephone ringer IC
C1, C4	1-μF capacitor
C2	470-pF capacitor
C3	4.7-μF, 50-V electrolytic capacitor
R1	3.9-kΩ, 1/2-W, 5% resistor
R2	220-Ω, 1/2-W, 5% resistor
R3	2.2-kΩ, 1/2-W, 5% resistor
BZ1	Piezoelectric buzzer

the ring voltage on the telephone line. An internal bridge rectifier within the MC34012 rectifies the supply voltage for use in the circuit. Resistor R1 and capacitor C1 act as a simple line voltage filter. In some cases these filtering components might not be needed, but it is vastly better to include them in the circuit and not need them, then to omit them and risk needing them. The exact values of these filter components are not too critical. I doubt that much would be gained by experimenting with alternate values for R1 and C1.

When a ring signal is received from the telephone line, the MC34012 drives a simple buzzer (BZ1). Many other output devices can be substituted in place of the buzzer. An oscillator circuit might produce a more pleasant ring sound.

The MC34012 is a two-frequency oscillator, continuously switching between a high frequency and a low frequency as long as the ring voltage is present on the telephone line. This gives a more distinctive, easier-to-hear ringing sound. The nominal switching rate for this circuit is 12.5 Hz.

The relevant oscillator frequencies in this circuit are determined by the components connected to pin 8 of the MC34012—resistor R2 and capacitor C2. Experiment with various values for these two components. I strongly recommend keeping the resistor value between 150 and 330 kΩ and the capacitor value between 400 and 2000 pF. This will permit base frequencies in a range from 1 kHz (1000 Hz) to 10 kHz (10,000 Hz).

There are three versions of the MC34012 available, each with a different set of nominal frequencies. The device type is indicated by a simple suffix number (1, 2, or 3). The MC34012-1 has a low frequency of 822 Hz and a high frequency of 1040 Hz. The low frequency for the MC34012-2 is 1664 Hz and the high frequency is 2080 Hz. Finally, for the MC34012-3 the frequencies are 416 Hz and 520 Hz.

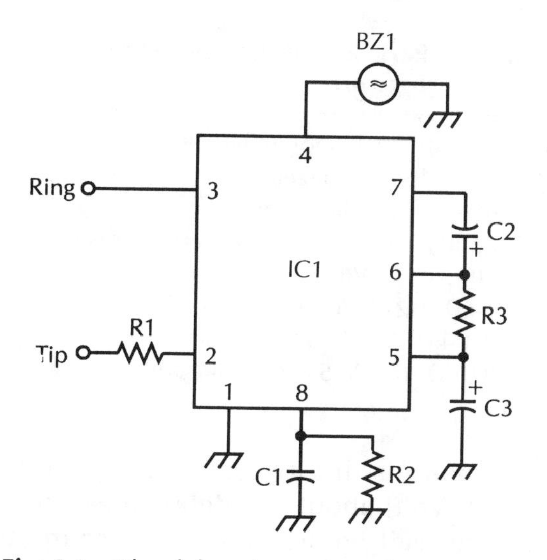

Fig. 4-6 *The delayed response ringer project.*

Notice that in all three cases, the low frequency and the high frequency are not harmonically related. Again, this makes for the most distinctive and audible sound quality.

Delayed response ringer

Figure 4-6 shows a rather unusual variation on the basic MC34012 ringer circuit described previously. When this circuit detects an incoming ring signal, it does not respond right away. Instead, there is a slight delay before the buzzer is sounded by the circuit.

A suitable parts list for this novel project is given as Table 4-3. Feel free to experiment with alternate component values in this circuit—especially resistors R1 and R3 and capacitor C2. The values of these three components determine the delay time for a given ring voltage. Unfortunately a neat design equation cannot be given for this delay function because of the complex way the ring signal voltage and the relevant component values interact. The values suggested in the parts list are a good starting point for further experimentation on your own. Roughly speaking, increasing any of the component values or decreasing the input (ring signal) voltage increases the delay period.

It probably wouldn't make much sense to use a delay of more than a few seconds. A caller is likely to hang up before you have a chance to respond to the delayed ring. On the other hand, this might be one way to weed out unwanted nuisance calls (like

Table 4-3 Parts list for the delayed response ringer project of Fig. 4-6.

IC1	MC34012 telephone ringer IC
BZ1	Piezoelectric buzzer
C1	680-pF capacitor
C2	22-μF, 250-V electrolytic capacitor (see text)
C3	10-μF, 250-V electrolytic capacitor
R1	1-kΩ, 1/2-W, 5% resistor (see text)
R2	220-kΩ, 1/2-W, 5% resistor
R3	820-Ω, 1/2-W, 5% resistor (see text)

telephone solicitors). Just inform your friends (and anyone else you'd want to call you) about the delay ahead of time. They'll know they need to wait before you'll answer the telephone. A nuisance caller, on the other hand, will probably give up and hang up.

This circuit can also be adapted for use in other, non-telephone applications. It can produce a delayed audio response to any ac input signal in the MC34012's acceptable voltage range (24 to 120 V).

If you use this project with your telephone, you probably won't want the delayed ring function on all the time. You can add a deactivate switch in parallel with capacitor C1. This is a simple SPST switch. When it is open, the delay is on. Closing the switch, however, defeats the delay function, and the circuit provides an instantaneous response to ring signals, just like any normal ringer circuit. Effectively, you decrease the value of capacitor C1 to zero by closing the switch.

Swept frequency ringer

For a very distinctive ringer, you can try the circuit shown in Fig. 4-7. The parts list for this project is given as Table 4-4. Once again, this ringer circuit is designed around the MC34012 telephone ringer IC.

Instead of a warble tone switching back and forth between two discrete frequencies (as in the basic MC34012 ringer circuit), the tone is swept smoothly over a continuous range of frequencies, starting at the lowest frequency and gliding up to the highest frequency. Then it jumps back down to the lowest frequency and starts over.

The sweep effect is accomplished by an ascending sawtooth

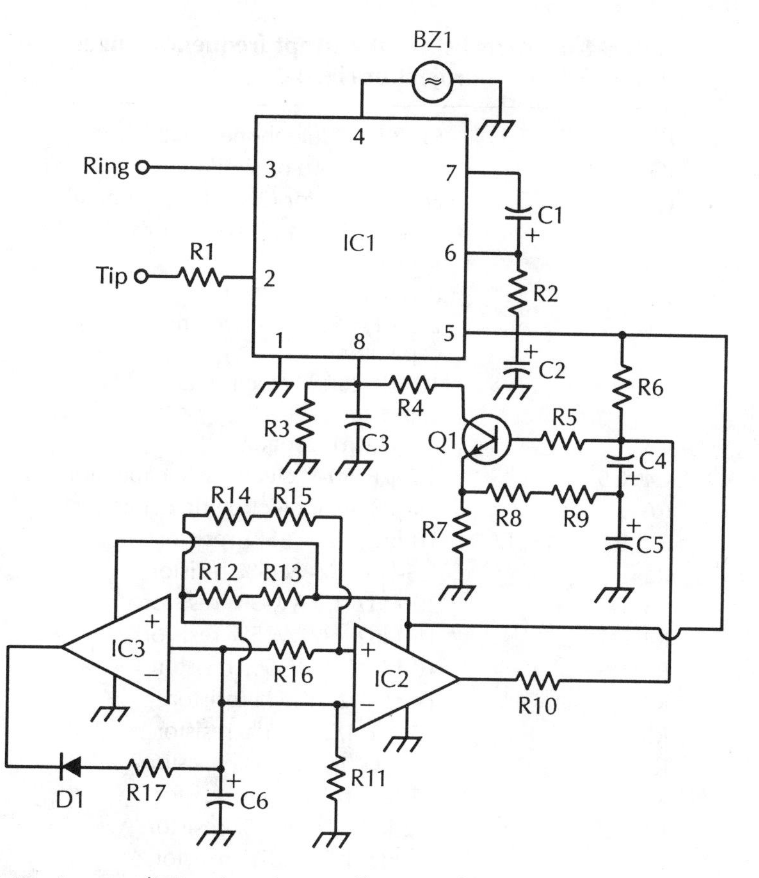

Fig. 4-7 *A very distinctive ringer effect can be achieved by sweeping the frequency through its range.*

wave oscillator made up of transistor Q1 and op amps IC2 and IC3. In effect, the transistor is gradually turned on as capacitors C4 and C5 charge up, which varies the effective resistance seen by pin 8 of IC1. Then the capacitors are quickly discharged and the cycle starts over.

You are encouraged to experiment with the various component values in this circuit. Probably the most worthwhile components for experimentation are capacitors C4, C5, and C6. These capacitors are the primary determinants of the depth and frequency of the sweep signal. Usually C4 and C5 should have equal values to keep the sweep waveform symmetrical. However, you might find the oddball effects of a nonsymmetrical sweep effect quite interesting.

Almost any low-power NPN transistor should work well for Q1. Similarly, almost any standard op amp ICs can be used for

Table 4-4 Parts list for the swept frequency ringer project of Fig. 4-7.

IC1	MC34012 telephone ringer IC
IC2, IC3	Op amp (748 or similar)
Q1	NPN transistor (2N3904 or similar)
D1	Diode (1N4148, 1N914, or similar)
BZ1	Piezoelectric buzzer
C1	2.2-μF, 250-V electrolytic capacitor
C2	10-μF, 250-V electrolytic capacitor
C3	0.0022-μF capacitor
C4, C5	4.7-μF, 50-V electrolytic capacitor
C6	15-μF, 50-V electrolytic capacitor
R1, R14, R15, R17	10-kΩ, 1/2-W, 5% resistor
R2	1.5-kΩ, 1/2-W, 5% resistor
R3	390-kΩ, 1/2-W, 5% resistor
R4	150-kΩ, 1/2-W, 5% resistor
R5, R6	10-kΩ, 1/2-W, 5% resistor
R7	1-kΩ, 1/2-W, 5% resistor
R8	47-kΩ, 1/2-W, 5% resistor
R9	3.9-kΩ, 1/2-W, 5% resistor
R10	47-kΩ, 1/2-W, 5% resistor
R11	4.2-kΩ, 1/2-W, 5% resistor
R12	15-kΩ, 1/2-W, 5% resistor
R13	27-kΩ, 1/2-W, 5% resistor
R16	2.2-MΩ, 1/2-W, 5% resistor

IC2 and IC3. There isn't much point in investing in high-grade, low-noise devices for an application like this. You might consider using a dual op amp such as the 747 or 1458 in place of the two separate op amp chips shown in the schematic. The only change you'll have to make (other than the pin numbers, of course) is that you'll need to make one set of power connections, instead of two. Notice that the op amps "steal" their operating power indirectly from the telephone lines. More specifically, they use the voltage at pin 5 of the MC34012 IC.

As usual, the actual frequency range generated for any given set of frequency components will depend on which version of the MC34012 you use. Remember, the MC34012-3 gives the lowest frequency range, the MC34012-2 gives the highest frequency range, and the MC34012-1 has a frequency range approximately midway between the other two. You might want to decrease the value of capacitor C1 with an MC34012-2 or increase it with a MC34012-3.

Warble-tone ringer

The ringer circuit shown in Fig. 4-8 produces a very unique and unmistakable warbling sound when the telephone rings. Unlike the last few projects presented in this chapter, this circuit does not require a dedicated ringer IC such as the MC34012. The effect is rather similar to that of the basic MC34012 ringer circuit.

A project like this is quite useful in an office or home with more than one telephone line. You can tell just by the sound which telephone (or line on a multiline telephone) is ringing. It can also be useful in other circumstances. For example, some

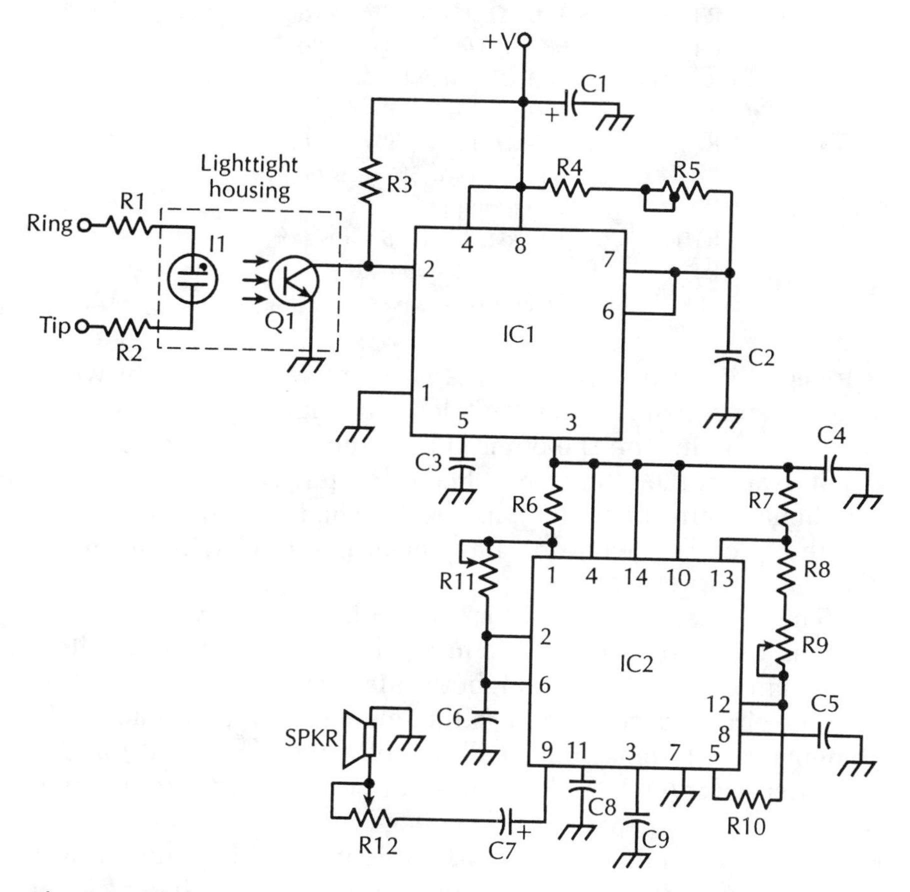

Fig. 4-8 *This ringer circuit produces a very unique and unmistakable warbling sound.*

Table 4-5 Parts list for the warble-tone ringer project of Fig. 4-8.

IC1	555 timer
IC2	556 dual timer
Q1	NPN phototransistor
I1	NE-2 neon lamp
C1, C4	250-μF, 35-V electrolytic capacitor
C2, C7	10-μF, 35-V electrolytic capacitor
C3, C8, C9	0.01-μF capacitor
C5	0.1-μF capacitor
C6	2.2-μF, 35-V electrolytic capacitor
R1	82-kΩ, 1/2-W, 5% resistor
R2	68-kΩ, 1/2-W, 5% resistor
R3	220-kΩ, 1/4-W, 5% resistor
R4	68-kΩ, 1/4-W, 5% resistor
R5, R11	500-kΩ trimpot
R6	2.2-kΩ, 1/4-W, 5% resistor
R7	4.7-kΩ, 1/4-W, 5% resistor
R8	1-kΩ, 1/4-W, 5% resistor
R9	10-Ω trimpot
R10	100-kΩ, 1/4-W, 5% resistor
R12	500-Ω potentiometer
SPKR	Small 8-Ω speaker

years ago I lived in a two-room apartment with very thin walls between apartments. I had my telephone, and my neighbor had his telephone just on the other side of the thin wall from mine. When I was in the other room, I couldn't tell whether it was my telephone ringing or my neighbor's. If I'd had a distinctive ringer, like this project, I wouldn't have been bothered with so many unnecessary interruptions.

The suggested parts list for this warble-tone ringer project is given as Table 4-5. You are encouraged to experiment with alternate component values throughout this circuit.

The circuit is connected to the telephone lines via a simple homemade optotransistor, made up of a neon lamp (I1) and a phototransistor (Q1), enclosed in some sort of lighttight housing. Ordinarily there is not enough voltage on the telephone line to light the neon lamp. (An LED would light up with a much lower voltage.) Only during the relatively high-voltage ringer signal will the neon lamp fire. Resistors R1 and R2 are included for current-limiting purposes.

The phototransistor (Q1) acts as a simple electronic switch, controlled by the neon lamp. When the lamp is dark (no ringer signal on the line), the phototransistor switch is off. However, when the lamp is lighted (ringer signal present on the line), the phototransistor switch is turned on, triggering IC1, a monostable multivibrator circuit.

The main body of this circuit is made up of three 555-type timers. You should have no trouble finding these ICs because they are one of the most popular and common ICs, and are likely to continue in their popularity for some time to come. Two of the 555 timers are contained in a single 556 dual timer IC (IC2). This makes the circuitry a little neater and more compact. You can substitute two separate 555s for the 556, if you prefer. Figure 4-9 shows the appropriate connections to make for this conver-

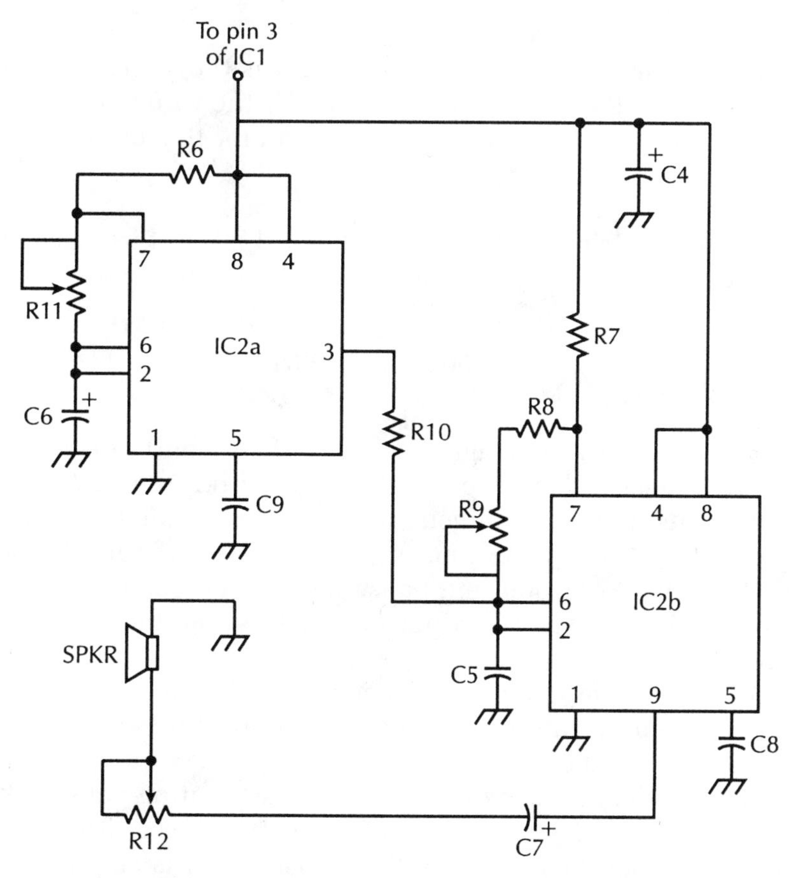

Fig. 4-9 *The warble-tone ringer project can be modified to use two separate 555 timer ICs instead of the 556 dual timer chip shown in Fig. 4-8.*

sion. This modification makes no difference in the actual operation of the project.

By using three timer stages in sequence, as in this project, a very distinctive warbling tone is generated. The first timer stage (IC1) is wired as a simple monostable multivibrator, or one-shot timer. When it is triggered by a ringer signal passing through the optoisolator (I1 and Q1), its output goes high for a predetermined period of time, determined by the values of resistors R4 and R5 and capacitor C2. This timer controls how long each burst of tone from the speaker will last. Resistor R5 is a trimpot to provide manual calibration of the tone duration. If you don't need such a control, you can combine R4 and R5 into a single fixed resistor of an appropriate value.

The standard timing equation for a 555 monostable multivibrator circuit is

$$T = 1.1RC$$

where T is the time period in seconds, R is the series combination of resistor R4 and trimpot R5, and C is the value of capacitor C2. You can rewrite the standard equation for this circuit as

$$T = 1.1 \times (\mathrm{R4} + \mathrm{R5}) \times \mathrm{C2}$$

Assuming trimpot R5 is set to the midpoint of its range, the time period will have a value of

$$\begin{aligned} \mathrm{T} &= 1.1 \times (47{,}000 + 250{,}000) \times 0.000\,01 \\ &= 1.1 \times 297{,}000 \times 0.000\,01 \\ &= 3.3 \text{ seconds} \end{aligned}$$

An ordinary ringer signal consists of 2-second pulses, separated by a 4-second silence. This circuit (with R5 set to the midpoint of its range) will generate 3.3-second bursts of tone, separated by 2.7-second pauses. The duration of each burst of tone (and the corresponding pause lengths) can be increased or decreased by adjusting trimpot R5. As the tone burst length is increased, the pause length is decreased.

With the component values suggested in the parts list, the tone burst length can be adjusted from a minimum of about 0.5 second (separated by 5.5-second pauses) to a maximum of a little over 6 seconds. In this case, there are no separating pauses—the tone is continuously sounded as long as the ringer signals keep coming through the telephone lines. Just as IC1 is about to time out, it is retriggered by the next burst of the ringer signal, restarting the time period from zero.

The output of IC1 controls the next timer stage (IC2a). Power is applied to the circuit only when the output of IC1 goes high—during the timing period. So the later half of the circuit can operate only when the first timer stage is triggered.

Basically the two timer stages in IC2 (IC2a and IC2b) are also monostable multivibrators, but they interact to serve as an astable multivibrator or rectangular-wave generator. Oversimplifying somewhat, trimpot R9 controls the output signal frequency, while trimpot R11 controls the duty cycle. *Duty cycle* refers to the proportion of the high and low portions of the rectangular wave. A few rectangular waves with differing duty cycles are shown in Fig. 4-10. The duty cycle of a rectangular wave determines the harmonic content of the generated tone. In other words, the timbral quality, or how the tone sound is affected by changing the duty cycle.

It is important to realize that the duty cycle and the frequency interact. If you change the duty cycle via trimpot R11, the signal frequency will be shifted, unless compensated for by a corresponding adjustment of trimpot R9 in the opposite direction.

If you don't want to bother with manually adjusting the duty cycle and/or frequency, you can eliminate one or both of these trimpots. You can combine resistor R8 and trimpot R9 into a single fixed resistor of an appropriate value. Similarly you can also (or instead) combine resistor R6 and trimpot R11 into a single fixed resistor of an appropriate value.

Capacitor C7 removes any possible dc component in the output signal before it can reach the speaker and possibly do some damage. Potentiometer R12 serves as a volume control. In most applications this should be a front-panel control, but you can use a trimpot if that suits your needs better. If you don't need a volume control for your intended application, you can replace this potentiometer with a fixed resistor. The higher the value of this component, the lower the volume of the produced sound will be. This volume-limiting resistor can be eliminated from the circuit to achieve the maximum possible output volume. If your application calls for an even higher volume—perhaps in a noisy factory environment, or something similar, you can eliminate the speaker and feed the output signal from IC2 (IC2b) into an appropriate audio amplifier system.

You can do quite a bit of experimentation with the component values in this circuit. I recommend hooking it up on a solderless breadboard to fiddle around with the resistor and

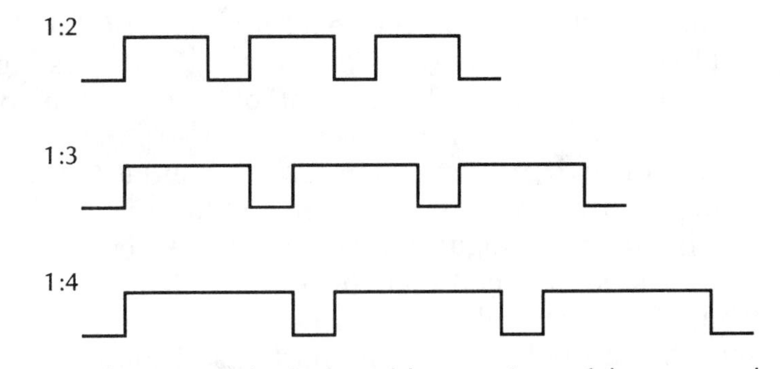

Fig. 4-10 The ratio of the high and low portions of the rectangular wave define the wave's duty cycle.

capacitor values throughout the circuit until you find the sound you like best. Some really interesting effects can be achieved by changing the value of resistor R10, which feeds the output of IC2a to the controlling input of IC2b. This will affect the warble quite drastically.

There isn't much point in experimenting with alternate values for capacitors C3, C8, and C9, however. In fact, you might be able to get away with eliminating these capacitors from the circuit altogether. These capacitors ensure the stability of each of the timer circuits. I consider such a stability capacitor to be cheap insurance against possible frustration anytime a 555 (or similar) timer chip is used without using the voltage control input. The exact value of the stability capacitor is not critical. Changing its value does not alter the operation of the circuit in any noticeable way.

I had a lot of fun working with this project, and I think you will too. Using this very distinctive sound for a telephone ringer can be a very practical solution to many annoying little problems.

Visual ringer

Have you ever missed a telephone call because there was too much noise and you didn't hear the telephone ring? Perhaps you were working with a power tool or some sort of machinery. Or maybe the kids were fighting, or you simply had the stereo cranked up.

Telephone ringers are designed to be as distinctive sounding and as attention getting as possible, but sometimes they just can't cut through existing noise in the environment. A visual ring indicator is very helpful in such cases. This project is also useful for people who are hard of hearing. When the telephone rings, a large LED flashes on and off at a frequency of about 5 Hz for a little over 4 seconds on each ring.

The schematic diagram for this visual ringer project is shown in Fig. 4-11. A suitable parts list for this circuit is given as Table 4-6.

This project uses the MC34012 telephone ringer IC. It also uses two 555 timer ICs. In this circuit, two separate timer ICs must be used. Do not use a 556 dual timer chip, because the output signal of timer IC1 provides the supply voltage for timer IC2. A dual 556 IC uses the same supply voltage for both timer sections, which will not work in this project.

The bottom portion of the schematic (IC3 and its associated components) should look familiar. This part of the circuit is essentially the same as the remote ringer project presented earlier

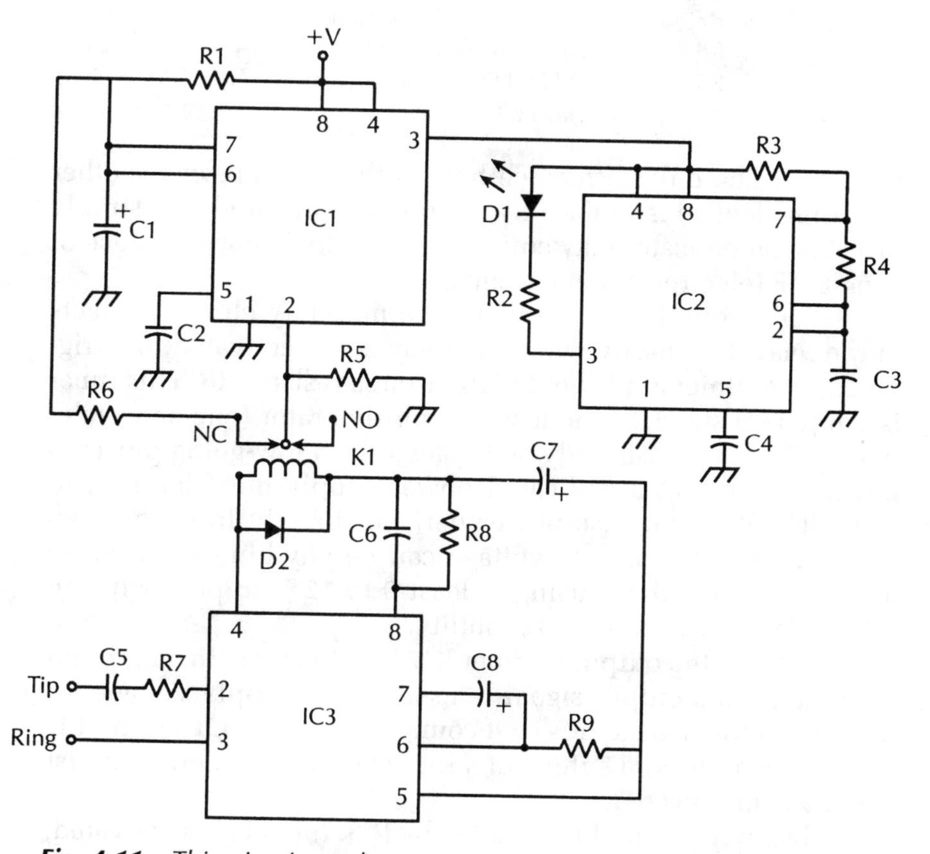

Fig. 4-11 *This circuit produces a visual rather than an audible response when the telephone rings.*

Table 4-6 Parts list for the visual ringer project of Fig. 4-11.

IC1, IC2	7555 or 555 timer
IC3	MC34012 telephone ringer IC
D1	LED
D2	Diode (1N4003 or similar)
C1	100-μF, 25-V electrolytic capacitor
C2, C4	0.01-μF capacitor
C3	0.47-μF capacitor
C5, C8	1-μF capacitor
C6	470-pF capacitor
C7	4.7-μF, 50-V electrolytic capacitor
R1	39-kΩ, 1/2-W, 5% resistor
R2, R6	47-Ω, 1/2-W, 5% resistor
R3	330-kΩ, 1/2-W, 5% resistor
R4	120-kΩ, 1/2-W, 5% resistor
R5	2.2-MΩ, 1/2-W, 5% resistor
R7	3.9-Ω, 1/2-W, 5% resistor
R8	220-kΩ, 1/2-W, 5% resistor
R9	2.2-kΩ, 1/2-W, 5% resistor
K1	Relay coil

(Fig. 4-5). The only difference here is that the buzzer (or other sounding device) is replaced with a relay coil. Diode D2 simply protects the delicate relay coil from possible damage because of back-EMF (electromagnetic frequency).

In this project, you are using the normally closed contacts of the relay. Ordinarily when the relay is not activated, the trigger input of timer IC1 is held high through resistor R6. This timer is wired as a standard monostable multivibrator (one-shot) telephone. This timer can only be triggered by a low-going pulse on pin 2. Notice that a dedicated power supply must be used to drive IC1. It cannot use power derived directly from the telephone lines. This supply voltage can be anything from +5 to +15 V. I would advise using at least 9 to 12 V to power IC1 in this project, for maximum reliability.

Normally the output of timer IC1 (pin 3) is low (near ground potential). This output signal is used as the supply voltage for the second timer stage (IC2). Of course, as long as IC1's output is low, IC2 cannot get a sufficient supply voltage to operate. It just sits there, unpowered.

When a ring signal is detected by IC3, the relay is activated, opening up the normally closed contacts. Now pin 2 of IC1 is pulled low through resistor R5. The one-shot timer is triggered, and its output (pin 3) goes high, with a voltage near the chip's supply voltage. This voltage is enough to power timer IC2, so it begins to operate. IC1's output will be held high for a period equal to

$$T = 1.1\text{C1R1}$$

Using the component values suggested in the parts list for this project, you get a time period of about

$$\begin{aligned} T &= 1.1 \times 0.000\,1 \times 39{,}000 \\ &= 4.29 \text{ seconds} \end{aligned}$$

After this timing period has expired, the output of IC1 goes low again, until it is retriggered by another ring signal. In the United States (and most of Europe), the telephone company's ring signal is comprised of a 2-second ring, followed by a 4-second pause before the next ring. Therefore the total ring signal cycle lasts 6 seconds. To increase visibility, in this project, you "ring" the flashing LED a little longer than normal on each cycle. The LED flashes for 4.29 seconds on each ring, with a 1.71-second pause between "rings."

The LED is controlled by the second timer stage (IC2), which is wired as a free-running astable multivibrator or rectangular-wave generator. As long as sufficent power is supplied to the IC, its output will continuously switch back and forth between high and low, turning the LED on and off. As you have already seen, this portion of the circuit gets power to operate only during the timing period of timer IC1. At all other times, this astable multivibrator circuit is simply turned off and the LED remains dark.

The flash rate, or frequency, is determined by the following formula:

$$F = 1.44/((\text{R3} + 2\text{R4})\text{C3})$$

Using the component values suggested in the parts list for this project, you get a flash frequency equal to

$$\begin{aligned} F &= 1.44/((330{,}000 + 2 \times 120{,}000) \times 0.000\,000\,47) \\ &= 1.44/((330{,}000 + 240{,}000) \times 0.000\,000\,47) \\ &= 1.44/(570{,}000 \times 0.000\,000\,47) \\ &= 1.44/0.2679 \\ &= 5.4 \text{ Hz} \end{aligned}$$

For some purposes an LED might be too small and not visible enough. You can modify the circuit by replacing LED D1 with

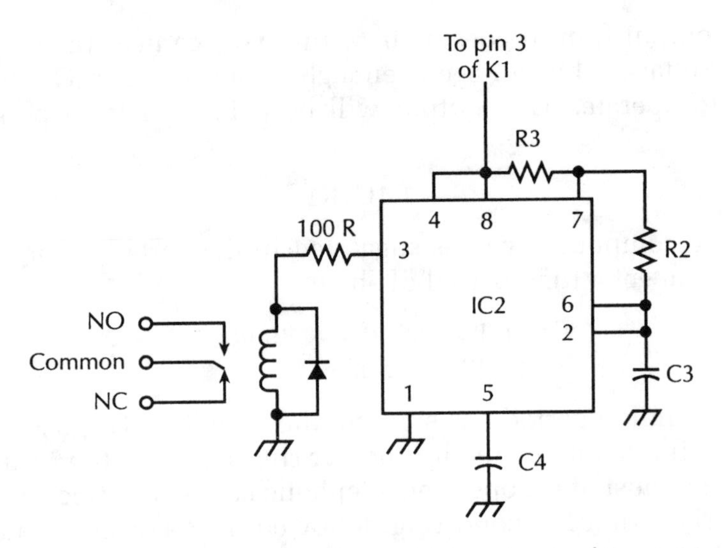

Fig. 4-12 A relay can be combined with the circuit of Fig. 4-11 to control a larger, brighter light source or almost any electrically operated device.

another relay that can control any light source (or whatever else) you'd like to use. You can even use a standard ac-powered lamp for a very visible display that is next to impossible to miss. The appropriate modifications to the project are illustrated in Fig. 4-12. Only the IC2 portion of the circuit is shown, because IC1 and IC3 and their associated components are completely unchanged for this modification to the visual ringer project.

Of course, this project does not limit you to a visual ring indication only. You can use this visual ringer circuit in parallel with an ordinary audible ringer circuit (including the ordinary ringer built into your telephone) for the best of both worlds. You'll be much less likely to miss a telephone call with both visual and audible indications when the phone rings.

Automatic ringer silencer

So far the projects presented in this chapter have been designed to let you know when you've got an incoming call, either through an audible ringer sound or through a visual indicator. But sometimes you don't want to be bothered by a ringing telephone, especially in the middle of the night. Many, but not all, telephones have a switch that permits you to silence the ringer. But it's all too easy to forget to turn the ringer off just before you go to bed. Perhaps more importantly, it is very easy to forget to turn the ringer back on in the morning, leaving you wondering why no one ever calls you anymore. Forgetting that the ringer on your telephone is turned off is a surefire way to miss potentially important telephone calls.

If your telephone is connected to a modular jack or an old four-prong jack, you can unplug it when you don't want to be disturbed by a ringing telephone. (If someone calls while the telephone is unplugged, the caller hears a normal ring-back signal—just as if no one was home to answer the ringing telephone.) It's still easy to forget, but its a little easier to remind yourself to plug your telephone back in—leave the cord and plug in plain sight, where you can't miss it. But its still a nuisance. Unless your telephone jack is ideally placed (and they rarely are), it will probably be rather inconvenient to unplug and re-plug the telephone.

This project will do the job for you automatically with a minimum of fuss and bother. The schematic diagram for this circuit is shown in Fig. 4-13. The parts list for this project is given as Table 4-7. Nothing is terribly critical here. Any pair of standard photoresistors can be used for R6 and R7. The resistance of these components varies with how much light strikes the surface of the device. The more light that hits the photoresistor, the lower the resistance, and vice versa. For a typical photoresistor, the resistance will be about 100 Ω when fully illuminated, and close to 3 MΩ (3,000,000 Ω) when fully darkened. This property is crucial for this project.

The telephone's plug must be modified slightly for this proj-

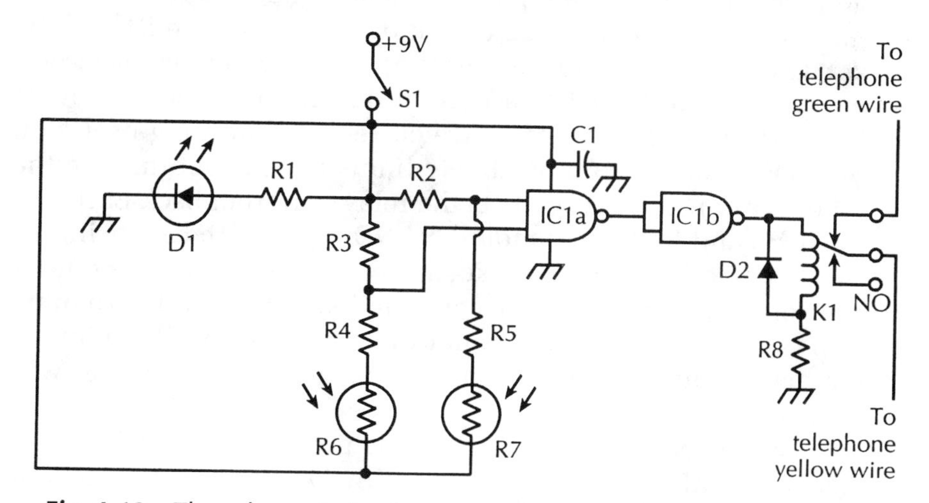

Fig. 4-13 *The schematic for the automatic ringer silencer project.*

Table 4-7 Parts list for the automatic ringer silencer project of Fig. 4-13.

IC1	CD4011 quad NAND gate
D1	LED
D2	Diode (IN4002 or similar)
K1	Small 9-V relay
S1	SPST switch
C1	0.01-μF capacitor
R1, R8	330-Ω, 1/4-W, 5% resistor
R2, R3, R4, R5	10-kΩ, 1/4-W, 5% resistor
R6, R7	Photoresistor

ect. Disconnect the telephone's yellow wire from the plug. Connect the green and red wires in the normal way. The telephone's yellow wire and a parallel wire connected to the telephone's green wire are hooked up to common and normally closed (NC) contacts of the relay. The relay's normally open (NO) contact is not used in this project.

When this circuit is turned off (by opening switch S1), the relay is deactivated because there is no power flowing through the circuit. The relay's common connection is shorted to the normally closed contact, shorting the telephone's yellow wire to the green wire. This is the way a telephone is usually wired within its plug. This connection permits the telephone to ring.

Now let's power up the circuit by closing switch S1 and see what happens. The LED D1 lights up to let you know the circuit is on-line. Resistor R1 is a current-limiting resistor to prevent the LED from drawing excessive current and burning itself out. Similarly resistor R8 protectively limits the current through the relay's coil. Diode D2 protects the relay coil from back EMF.

IC1a and IC1b are standard NAND gates. IC1b has its inputs shorted together so it acts like an inverter, reversing the output state of IC1a. This makes the combination of IC1a and IC1b function as an AND gate. The output is high if and only if both inputs are high. If either or both of the inputs are low, the output will also be low.

Two NAND gates are used in this project instead of a single AND gate, simply for convenience. NAND gates are easier to find and generally less expensive than AND gates, and because there

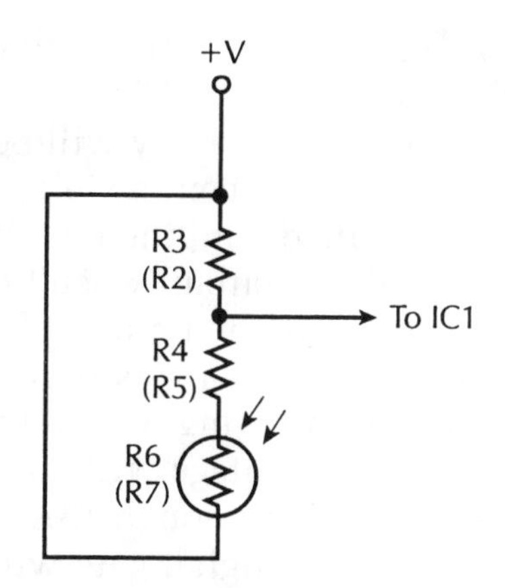

Fig. 4-14 *One of the two resistive voltage divider networks connected to the inputs of IC1 in Fig. 4-13.*

are four gates in a single IC, this approach doesn't increase the parts count of the project.

Capacitor C1 protects the CMOS IC from any possible noise spikes on the supply voltage line. It's exact value is not particularly critical.

Notice that there are two resistive voltage divider networks connected to the two inputs of IC1a. These voltage divider networks are identical. To make it easier to see, the first voltage divider string is shown all by itself in Fig. 4-14. This voltage divider network is made up of resistors R3 and R4 and photoresistor R6. Notice that this voltage divider string is connected at either end to the supply voltage. The junction of resistors R3 and R4 is connected to one of the inputs to IC1a.

For the time being, let's pretend that photoresistor R6 doesn't exist at all in the circuit. This means the voltage divider string is made up of two equal-value resistors—R3 and R4. Because these resistances are equal, the voltage drop across them must also be equal. This means the voltage at their junction (the input to IC1a) is zero.

Including photoresistor R6 in the voltage divider network throws this perfect balance off, at least a little. The input to IC1a will never be zero, but that doesn't really matter.

Let's first assume the area is brightly lit, so the photoresistor's surface is fully illuminated. In this case, R6's value will only be about 100 Ω, so (ignoring all component tolerances in the circuit) the series combination of R4 and R6 will be 10,100 Ω, which isn't all that different from the 10,000 Ω of R3. The input voltage applied to IC1a will be very close to zero. It will be interpreted as a logic low by the digital gate. Remember, whenever any of the inputs to an AND gate is low, the output must also be low. This means the relay will be deactivated. The normally closed connection between the telephone's yellow and green wires will be maintained, and the telephone will function completely normally, as if this circuit wasn't hooked up at all.

So far we haven't accomplished much of anything. But let's consider what happens as the area gets dark. The resistance of the photoresistor will rise, throwing off the balance of the voltage divider string. A positive voltage will appear at the input of IC1a. If this voltage is positive enough (resulting from a large enough resistance for R6), the digital gate will see a logic high. If both inputs are high, then IC1b's output will also be high, activating the relay. This disconnects the normally closed connection between the telephone's green and yellow wires, disabling the unit's ringer circuit. The rest of the telephone will function normally. Outgoing calls will not be affected by this circuit in any way. Only the ringer will be silenced. If someone calls while the relay is activated, they hear the normal ring-back signal of an unanswered telephone, but you aren't disturbed.

A second identical voltage divider network is made up of resistors R2, R5, and R7. Both photoresistors must be sufficiently darkened to activate the relay and turn off the ringer function. This makes the system much more reliable. If you use just a single photoresistor as the light sensor, the circuit could become confused and falsely activated by an inadvertent shadow.

You can see how this circuit functions automatically. During the day, the photoresistors are illuminated, keeping the resistance low and the relay deactivated. The telephone will react to any incoming ring signals in the normal way. At night, when you go to bed and all the lights are off, the photoresistors are dark and the relay is activated, disabling the telephone's ringer.

If you think you might receive an important late-night call—perhaps if a relative is seriously ill or your teenager is out on a late date—you can turn off the automatic ringer silencer circuit via switch S1, and your telephone will operate normally.

It might seem that this project won't be of much help if you want to disable the ringer during the day, but its always possible to come up with a solution using a little creativity. Just cover up the photoresistors to block them from any light. Just remember to unblock the sensors later when you want to reactivate your telephone's ringer.

This project is not as obvious as some, and that might be why it's one of my personal favorites.

❖ 5 Hold buttons

A valuable and useful feature found on most business telephones is the hold button. By pressing this button, you can temporarily hang up the phone for whatever reason without disconnecting the call. You can then pick up the receiver (perhaps on a different extension) and resume the conversation.

Hold buttons are not offered on most telephones designed for home use—there are a few, but they are relatively uncommon. This is a shame because a hold button can be implemented in any telephone without a lot of expensive circuitry. And it is such a useful function.

Have you ever needed to talk to someone in the room without letting the person on the telephone hear what you are saying? You can cover the mouthpiece, but that usually isn't a very effective or reliable way to avoid being overheard. With a hold button, you can prevent the caller from listening to your local conversation.

If you have more than one telephone in your home, you have undoubtedly answered an incoming call on one phone, then for some reason needed to change to another phone elsewhere in the house. You can leave the first phone off the hook, run to the desired extension, pick up its receiver, and tell your caller, "Hang on a minute." Then you leave the second phone off the hook while you run back to the first phone to hang it up, returning to the second phone to complete the conversation. If there are others in your household, it can become even more of an adventure. Someone might see one of the telephones off the hook and "helpfully" hang it up.

You could just leave the first telephone off the hook while you complete your conversation on the second phone, but this is also far from ideal. It's all too easy to forget to hang up the first phone after you have completed the call. You could miss another (possibly important) incoming call because you left the first telephone off the hook. Also leaving the first phone off the hook is an open invitation to any potential eavesdroppers in your household. In many cases, having two telephones on the same line simultaneously off the hook reduces the volume and increases the noise, which is hardly desirable.

A hold button is a perfect solution to such problems. You can answer the first telephone as before. When you realize you need to switch phones, you press the hold button, and hang up the first phone. Now you can go to the second phone, pick up the handset, and resume your conversation. You don't need to bother with the first telephone at all now. It's already been hung up.

The telephone lines carry a voltage between the ring (red) and tip (green) connections to your telephone. When the handset is on the hook, a switch is opened in the telephone (as shown in Fig. 5-1), breaking the connection between the ring and tip lines. The equipment at the telephone company senses the voltage is not coming through, so it breaks any call connection and considers the line open to receive new incoming calls.

When the handset is lifted from the hook, the switch inside

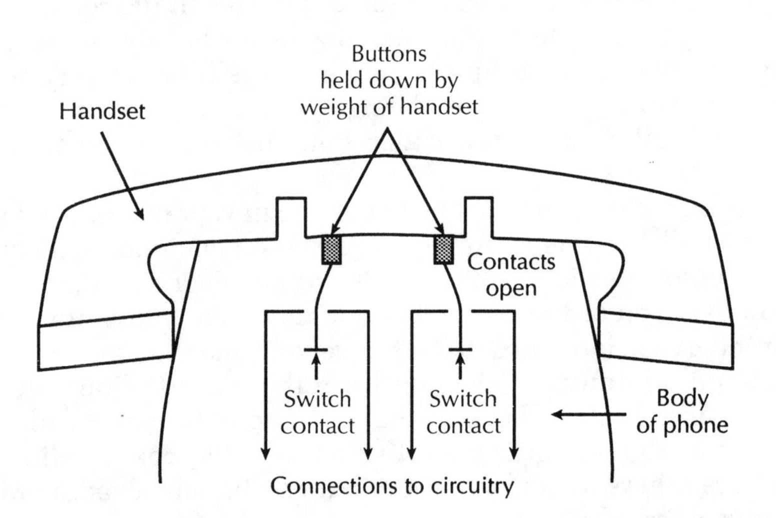

Fig. 5-1 *When the handset is on the hook, a switch inside the telephone is opened.*

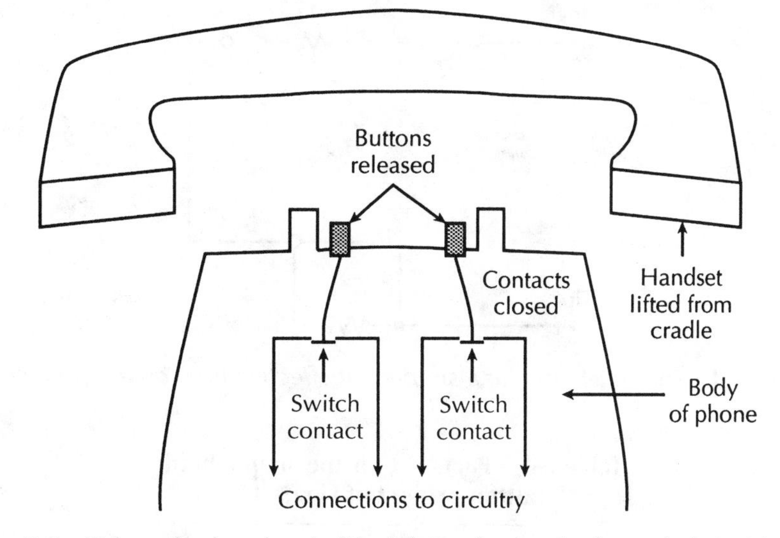

Fig. 5-2 *When the handset is lifted from the hook, the switch inside the telephone is closed, completing the circuit between the ring and tip lines.*

the telephone is closed, completing the circuit between the ring and tip lines, as illustrated in Fig. 5-2. The telephone company's equipment senses that the voltage is passing through your telephone, indicating a call in progress. Your telephone line is treated as busy.

A hold button's job is to fool the telephone company's equipment into thinking the handset is still off the hook, even though you have actually hung up your telephone. As you might have guessed, this is simply a matter of electrically completing the circuit between the ring and tip leads so the telephone company's equipment sees the appropriate off-hook voltage.

Hold button circuit

A simple, but effective hold button circuit is shown in Fig. 5-3. A suitable parts list for this project is given as Table 5-1.

This hold button is easy to use. Simply hold push-button switch S1 closed as you hang up the telephone. This turns on the SCR (Q1). The LED (D1) lights up, indicating the hold function has been activated.

Normally the SCR looks like an open circuit, so this circuit doesn't do anything. When the SCR is not conducting, no current flows through the LED, and it remains dark. Electrically the hold

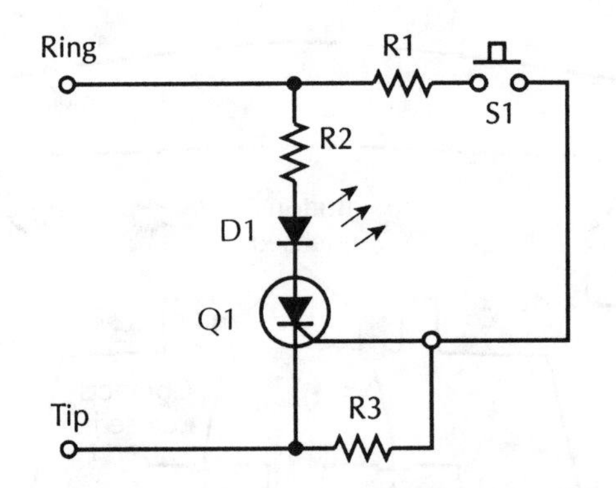

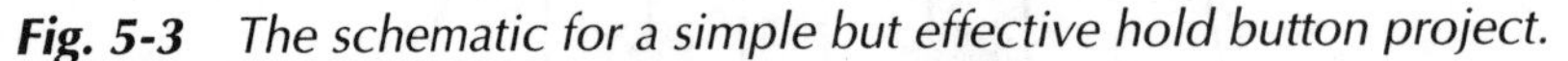

Fig. 5-3 *The schematic for a simple but effective hold button project.*

Table 5-1 Parts list for the simple hold button project of Fig. 5-3.

Q1	Low-power SCR, rated for at least 200 V
D1	LED
R1	100-kΩ, 1/2-W, 5% resistor
R2	2.2-kΩ, 1/2-W, 5% resistor
R3	47-kΩ, 1/2-W, 5% resistor
S1	Normally open SPST push-button switch

circuit might as well not be connected to the telephone line at this point. It makes no difference.

Closing switch S1 feeds the incoming ring voltage (through resistor R1) to the gate of the SCR. Once it has been turned on, the SCR conducts current, permitting the LED to light up. The conducting SCR also feeds the ring voltage through to the tip wire, even though the telephone's handset is on the hook. The call will not be disconnected.

When the telephone's handset (or another extension) is taken off the hook, you effectively have two resistances in parallel, as illustrated in Fig. 5-4. This reduces the voltage applied across the SCR, causing it to turn itself off. The telephone now functions normally for the remainder of the call. When the handset is hung up, the call will be disconnected, unless S1 is held closed again.

On some telephone lines, you might need to do a little experimentation to find the most reliable values for resistor R2, and possibly resistor R3. Resistor R1 limits the turn-on current permitted to flow through the gate of the SCR. The exact value of this resistor is not particularly critical in most applications.

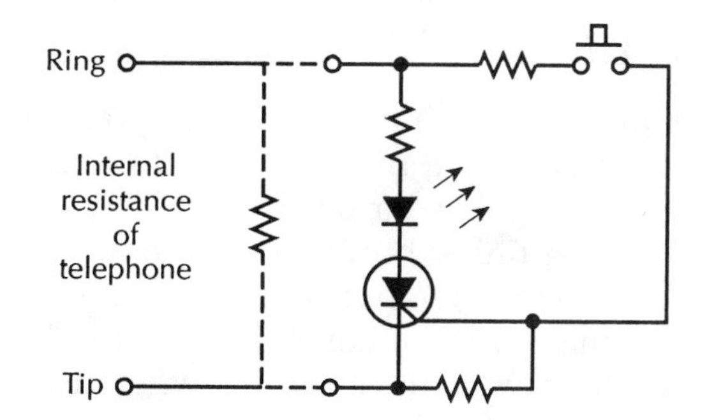

Fig. 5-4 *When the telephone's handset is taken off the hook, you effectively have two resistances in parallel, reducing the voltage applied across the SCR, causing it to turn itself off.*

The current through the LED and the SCR is limited by resistor R2. This resistance is moderately critical, and must be selected so that picking up an extension telephone's handset (or the original handset) drops the current flow through the SCR enough to ensure turn-off. The 2.2-kΩ resistor recommended in the parts list is a good starting point. If the operation of your hold button circuit is unreliable, try experimenting with slightly higher or lower resistances for this component. No lasting harm will be done if this resistor value is off a little from its ideal value—the project just won't work properly until R2's resistance is correct. Many variables can affect the ideal resistance, so I can't give you a hard and fast resistor value that will work in all cases. Finally, resistor R3 prevents the hold circuit from accidentally activating itself when it is not desired (possibly because of a noise spike).

The SCR used is not too critical. It must be able to withstand at least 140 V, which is the maximum ring voltage that will appear across normal telephone lines. A 200-V SCR is the nearest standard value, and it will offer plenty of headroom to protect the device against most noise spikes that might appear on the telephone lines.

This is certainly a very simple project. Even using all new components, it shouldn't cost more than $2 or $3 to build this circuit. Because it is so small, you might consider installing the circuit into the body of each of your telephones. (I am assuming you legally own the telephones in question.) Or if you prefer, you can construct the project as a separate stand-alone unit connected to the telephone lines via a Y-jack.

Music on hold

A simple telephone hold button like the preceding project is functional, but it is a little inconsiderate to your caller. While they are on hold, they have a dead line. They don't hear anything. Besides being rather boring, this can also cause concern that the call has been inadvertently disconnected. The caller has no way of knowing until you come back on the line or the dial tone clicks in. This is why most offices play music on hold.

Figure 5-5 shows a simple, inexpensive circuit that permits you to play music for a caller while you have them on hold. A suitable parts list for this project appears as Table 5-2.

The audio input can come from a radio or a tape player. You certainly aren't limited to the rather bland "elevator music" most professional music-on-hold systems employ. You can use whatever type of music you prefer, or you can play a tape of a spoken message, or any sound at all.

Ordinarily the SCR is off, so the audio input signal cannot reach the telephone lines. Holding momentary switch S1 closed as you hang up the phone activates the hold mode. The SCR is turned on. Its conduction simulates an off-hook handset, as far as the telephone lines are concerned. It also activates relay K1, permitting the audio signal to pass through to the telephone lines.

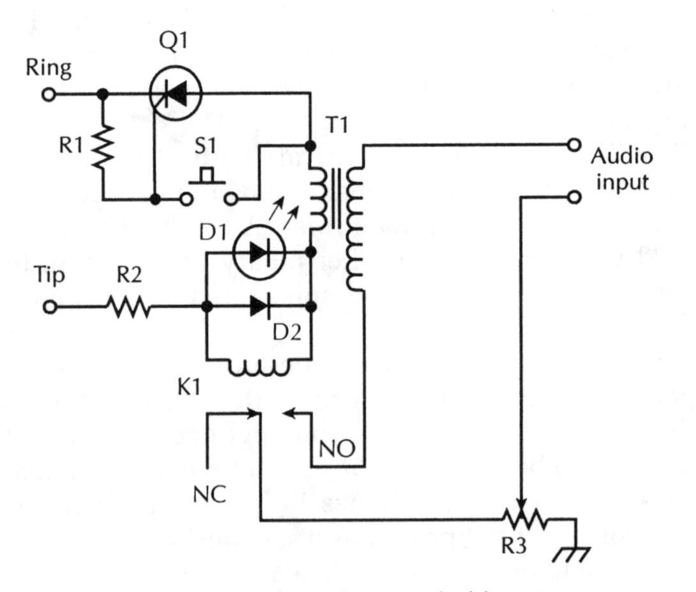

Fig. 5-5 *An inexpensive music-on-hold project.*

Table 5-2 Parts list for the music-on-hold project of Fig. 5-5.

Q1	Low-power SCR, rated for at least 200 V
D1	LED
D2	Diode (1N4003 or similar)
K1	Small relay
T1	Small impedance-matching transformer, 1000:8 Ω
S1	Normally open SPST push-button switch
R1, R2	470-kΩ, 1/2-W, 5% resistor
R3	1-kΩ potentiometer

Transformer T1 is an impedance-matching transformer, selected to match the output impedance of the audio signal source. For example, if you are using a tape recorder with a 1000-Ω earphone jack, you would use a 1000:8-Ω transformer, as suggested in the parts list. The 8-Ω secondary is connected to the telephone side, to the SCR (Q1) and the relay coil (K1).

When the relay is deactivated (the hold function is not in effect), the audio input circuit to the transformer's primary is broken. When the relay is activated by pushing the hold button (S1), the normally open contacts close, completing the audio input circuit. LED D2 also lights up as an indicator that the hold function is activated.

Potentiometer R3 is an optional volume control to set the level of the audio signal fed into the telephone lines. In many cases, you might want to omit this control and just use the volume control built into the audio signal source. In this case, simply eliminate R3 from the circuit. You don't need to replace the potentiometer with a fixed resistor.

If you are using a portable cassette tape recorder as your audio signal source, you probably won't want to keep the tape playing continuously. You could manually hit the recorder's play button whenever you activate the hold function, but this is a rather inelegant approach. And it is all too easy to forget.

Fortunately it is usually easy to automate the process with most portable cassette recorders. With few exceptions, such recorders have jacks for an external microphone. Most cassette recorder microphones have a simple remote control switch built in. Two jack pins are plugged into the recorder. One, a miniplug, carries the audio signal from the microphone. You are not interested in this plug. The other is a simple make-break switch

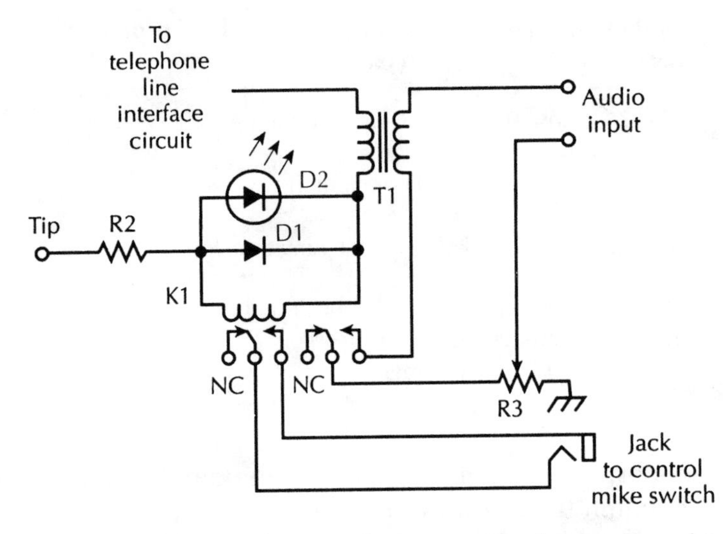

Fig. 5-6 *A second set of relay switch contacts can be connected to a microplug to simulate a remote microphone switch.*

through a microplug. When this remote switch is off (open), the recorder is deactivated. Its power source is internally disconnected. Closing the remote switch permits the recorder to function normally.

To modify this music-on-hold project to utilize a cassette recorder's remote switch function, you must use a DPDT (or normally open DPST) relay. The second set of relay switch contacts are connected to a microplug to simulate the remote microphone switch, as illustrated in Fig. 5-6.

6
Dialing projects

Dialing is an important function in the telephone system. This is how one telephone can be connected to any other in the worldwide telephone system.

Originally I intended to include some automatic dialer projects in this chapter. An automatic dialer includes an electronic memory bank that can store frequently dialed telephone numbers. Once stored, any of these numbers can be dialed by simply pushing one or two buttons. The user doesn't have to dial the entire number to place a call. This can be very convenient and a real time-saver. The advantage isn't so much that you only have to press one or two buttons instead of seven (or eleven for long distance) each time you place a call, although this is nice. The chief advantage of an automatic dialer, as I see it, is that it removes the need for you to remember or look up a number every time you call it. I don't know about you, but I've never found telephone numbers particularly easy to memorize.

An automatic dialer seemed like a very natural electronic project. However, when I tried to design some projects along these lines, I ran into problems. Many commercially manufactured dialers are available, increasingly at very reasonable prices. I've seen several for under $40. Mass-production construction methods and purchasing components in bulk permits the commercial manufacturers to be very competitive. I've found that building a home-brew automatic dialer project ends up costing more than a comparable commercially manufactured unit with similar features. Therefore, there really wouldn't be much point in going to the trouble of building your own.

But there are still some useful dialing-related projects that are worth your while. If you have an old-style rotary telephone or a push-button telephone with pulse dialing, a DTMF dialer circuit can permit you to take advantage of special functions available only with Touch-Tone® systems. More and more businesses use computerized telephone systems, which recognize DTMF signals as controlling input. These systems don't work with rotary or pulse dialing. Even if your local telephone line doesn't accept Touch-Tone® dialing for placing calls, it can still be very useful to have a DTMF tone source available.

DTMF tones can be used for many control purposes, even outside the realm of telephones. They are used in many remote control and communications applications. The projects in this chapter include a couple DTMF generators or encoders and a DTMF decoder (for recognizing received DTMF signals). There is also a one-number automatic dialer project that can literally be a lifesaver in some emergency situations.

DTMF audio generator

Dual-tone multifrequency signals are used in Touch-Tone® telephone dialing. They can also be used for many other purposes. Almost any electronic device or function can be controlled by these unambiguous audio tones.

A small portable DTMF audio generator can be used with a rotary or pulse-only telephone to provide Touch-Tone® signals. Simply hold the speaker of the generator close to the mouthpiece of the telephone's handset. The telephone company's central switching office will "hear" the DTMF tones on the line. The system doesn't care about the source of the tones, as long as the frequencies are correct.

Another application for a small portable DTMF audio generator is as an audible-signal remote control unit. It can also be used for coded signaling in some specialized applications.

A circuit for a practical DTMF audio generator project is shown in Fig. 6-1. A suitable parts list for this project appears as Table 6-1.

This circuit is designed with two different types of outputs—audio tones from a small speaker, or a relay's switch contacts. If your application needs only one of these outputs, there isn't much point in spending money for the components for the unused output. If you want to eliminate the relay output, for a DTMF generator with audio (speaker) output only, you can omit

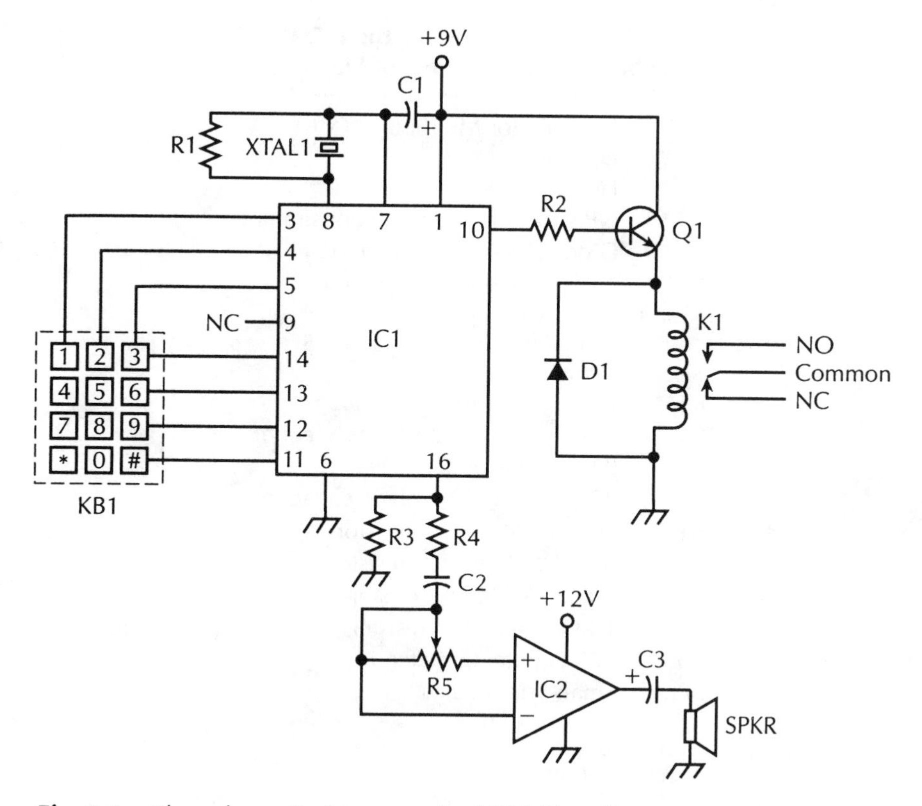

Fig. 6-1 *The schematic for a practical DTMF audio generator project.*

all of the components in the parts list marked with two asterisks (**). On the other hand, if you intend to use the relay output, but not the audio (speaker) output, you should omit the components in the parts list marked with a single asterisk (*).

Potentiometer R5 is a volume control. If your application doesn't require manually adjustable volume, you can replace this potentiometer with a fixed resistor. The higher this resistance is, the lower the output volume will be. I recommend keeping its value between 10 and 100 kΩ.

Notice that different supply voltages are used for IC1 and IC2. IC1 works best with a +9-V supply, while IC2 calls for +12 V. Do you need two separate power supply circuits for this project? No. A simple voltage divider network can be used to derive the lower supply voltage (+9 V) from the higher supply voltage (+12 V), as illustrated in Fig. 6-2. The value of resistor R6 should be one-third the value of resistor R7 to get a 9-V output at the center node with a 12-V input. I have found that using a 22-kΩ resistor for R6 and a 68-kΩ resistor for R7 works well. Notice

Table 6-1 Parts list for the DTMF audio generator project of Fig. 6-1.

IC1	MK5085 (or MK5086) DTMF generator
IC2*	LM380 audio amplifier
Q1**	NPN transistor (2N3904 or similar)
D1**	Diode (1N4003 or similar)
KB1	12-key telephone keypad
XTAL1	3.58-MHz crystal (actual frequency—3.579545 MHz)
K1**	Relay to suit load
C1	10-μF, 25-V electrolytic capacitor
C2*	0.01-μF capacitor
C3*	47-μF, 25-V electrolytic capacitor
R1	10-kΩ, 1/4-W, 5% resistor
R2**	8.2-kΩ, 1/4-W, 5% resistor
R3*	1-kΩ, 1/4-W, 5% resistor
R4*	1-kΩ, 1/4-W, 5% resistor
R5*	100-Ω potentiometer
SPKR*	Small 8-Ω speaker

*audio output
** relay output

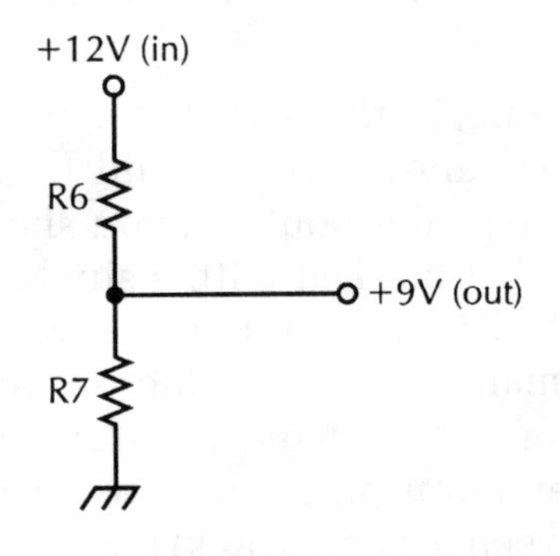

Fig. 6-2 A simple voltage divider network can be used to derive the lower supply voltage (+ 9 V) from the higher supply voltage (+ 12 V).

that the resistance ratios don't have to be exact. Nominally R6 is 33.33 percent of R7. Using these component values, R6 is actually 32.24 percent of R7. The component tolerances of the resistors account for at least this much error, so these resistance values are certainly close enough.

IC1 is critical in this project. Make sure you have a source for the MK5085 (or MK5086) DTMF generator IC before beginning the project. If you can't find this chip, you won't be able to use this circuit.

Several other DTMF generator ICs are available, but they usually have differing pin-outs and require somewhat different support circuitry. No matter what chip I picked, I ran into the same potential limitations. Trying to duplicate the DTMF generator's internal circuitry with more general-purpose components is theoretically possible, of course, but it is wildly impractical.

Almost any low-power NPN transistor will work well for Q1. This transistor functions as a simple electronic switch, and there are no special requirements for its operating parameters. If your application happens to call for a very large relay, you might need to use a heftier transistor. An even better approach is to cascade two relays. A small relay can be used to control a second, larger relay (with its own power source).

You should have no difficulty finding the crystal (XTAL1) used in this project. It is a standard 3.58-MHz crystal, probably the most commonly available frequency for crystals. This is the same frequency used for the color burst signal in television sets and many other applications. Actually, 3.58 MHz is a rounded-off value. The actual frequency of this crystal is 3.579545 MHz. Most crystals are marked with the more convenient 3.58 MHz, but don't be confused if you come across one marked with the more accurate extended value.

KB1 is a twelve-button matrixed keypad. These devices are available new, or you can find them at bargain prices from a variety of electronics surplus dealers. You could also take the keypad from a broken push-button telephone.

In some cases, you might find a four-by-four keypad (sixteen keys) instead of the standard three-by-four keypad (twelve keys). If you use a sixteen-key unit, you can simply leave the extra column disconnected. But in some control applications, it might be useful to have the four extra keys operational. Simply connect the lead of the extra column to pin 9 of IC1, which is included for exactly this purpose.

All of the other components used in this project are fairly standard. You should have no trouble finding any of them.

Alternate DTMF encoder

If you can't locate an MK5085 or other suitable DTMF generator IC, you're not completely out of luck. This project is a slightly cruder but still functional DTMF encoder, using only standard, commonly available components. Specifically, the tone encoding is done by three 555 timer chips. Except perhaps for op amps, this is probably the most popular and common type of IC in modern electronics.

In this particular application, I recommend using 7555 timer chips. The 7555 is a complementary metal-oxide semiconductor (CMOS) version of the standard 555. It is slightly more accurate and consumes less power. The 7555 and the 555 are pin-for-pin compatible. No changes need to be made in the circuitry to substitute one device for the other.

Another possibility is to use a 556 dual timer chip in place of two separate timers. I recommend combining IC1 and IC2 if you choose to do this. The 556 is the exact equivalent of two 555 chips in a single, convenient housing. Only the power supply connections are common to both sections of the 556. Otherwise the two timer sections are fully independent in all of their functions. No changes must be made in the circuitry to substitute a 556 dual timer for two 555 timers, but you must correct the pin numbers. For your convenience, the pin-out diagram for the 555 (and 7555) timer IC is shown in Fig. 6-3, and the 556 dual timer's pin-out diagram is shown in Fig. 6-4. The pin functions for these two chips are compared in Table 6-2.

By not using a dedicated DTMF generator IC, the project is necessarily complicated. To avoid making the details of the schematic diagram too small to see, the circuitry for this project is broken up into three sections. The main encoder circuitry is shown in Fig. 6-5, the relay output circuitry is illustrated in

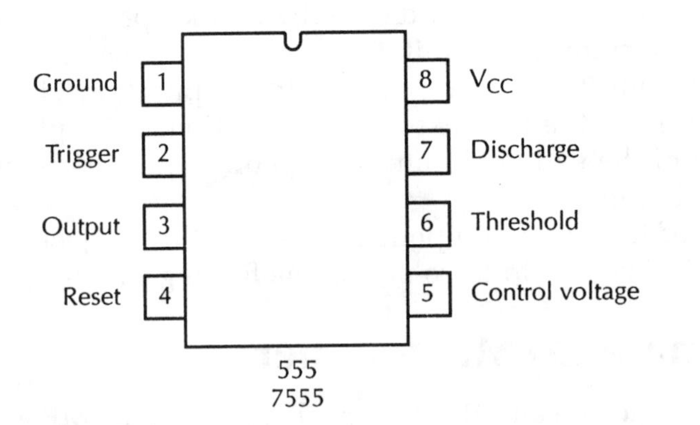

Fig. 6-3 *The pin-out diagram for the 555 (and 7555) timer IC.*

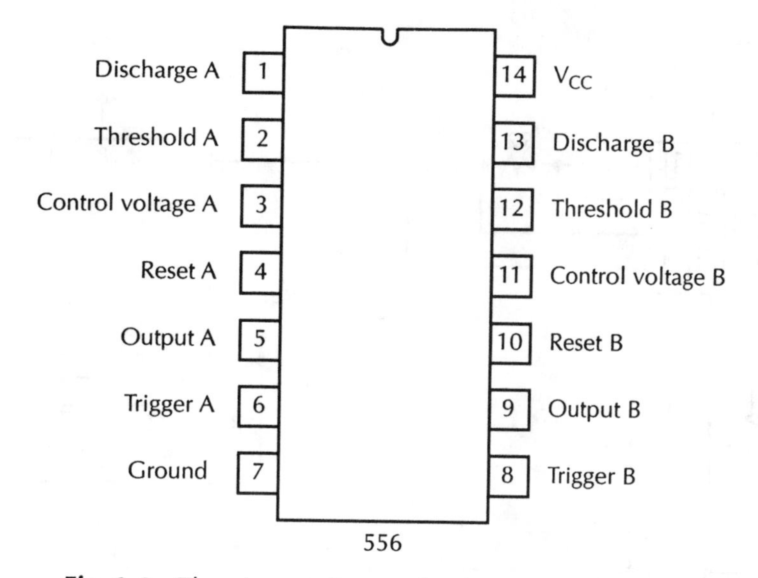

Fig. 6-4 *The pin-out diagram for the 556 dual timer IC.*

Table 6-2 Comparison of pin-out for the 556 dual timer and 555 timer ICs.

Function	556 dual timer	555 timer A	555 timer B
Discharge A	1	7	—
Threshold A	2	6	—
Control voltage A	3	5	—
Reset A	4	4	—
Output A	5	3	—
Trigger A	6	2	—
Ground	7	1	1
Trigger B	8	—	2
Output B	9	—	3
Reset B	10	—	4
Control voltage B	11	—	5
Threshold B	12	—	6
Discharge B	13	—	7
Vcc	14	8	8

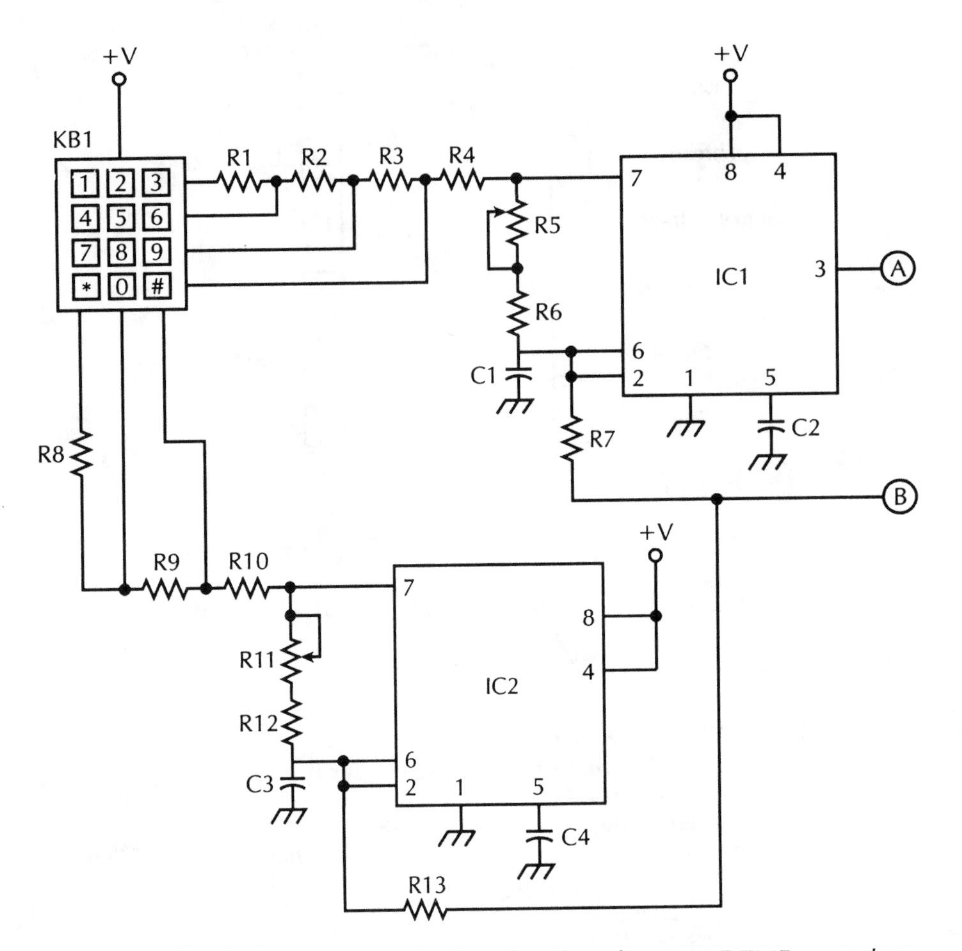

Fig. 6-5 *The main encoder circuitry for the alternate DTMF encoder project.*

Fig. 6-6, and the audio output circuitry appears as Fig. 6-7. The complete parts list for the entire project is given as Table 6-3.

All points in the three-part schematic marked V+ should be connected to the same power source. This can be anything within the acceptable supply voltage range for the 555 timer. I suggest using a supply voltage between +9 and +12 V. A standard 9-V transistor radio battery would be a fine power supply in a portable version of this project. An alkaline battery will offer the best and most reliable operation. Carbon-zinc batteries are likely to wear down too quickly to be truly practical. The points marked A and B should be connected to their matching points in the other sections of the schematic.

Notice that the keypad used in this project must be a somewhat specialized type that accepts an external supply voltage.

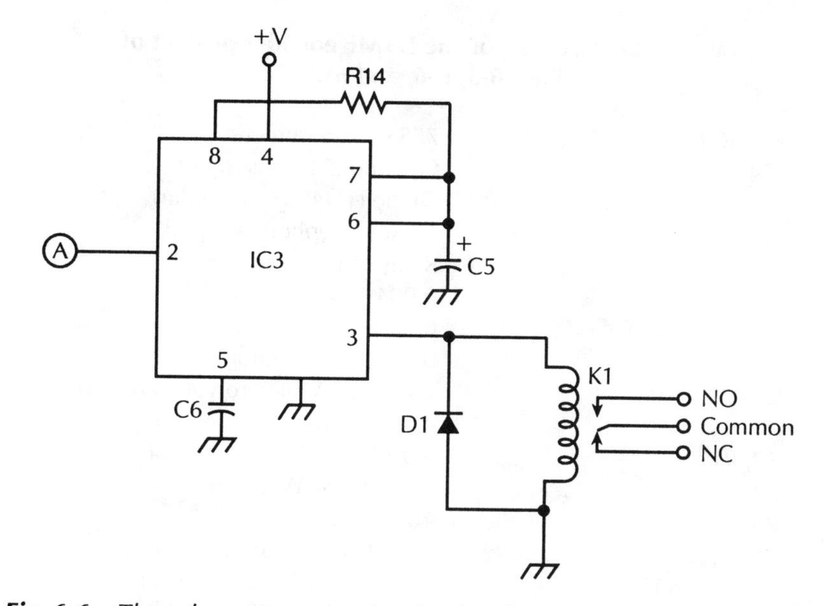

Fig. 6-6 The relay output circuitry for the alternate DTMF encoder project.

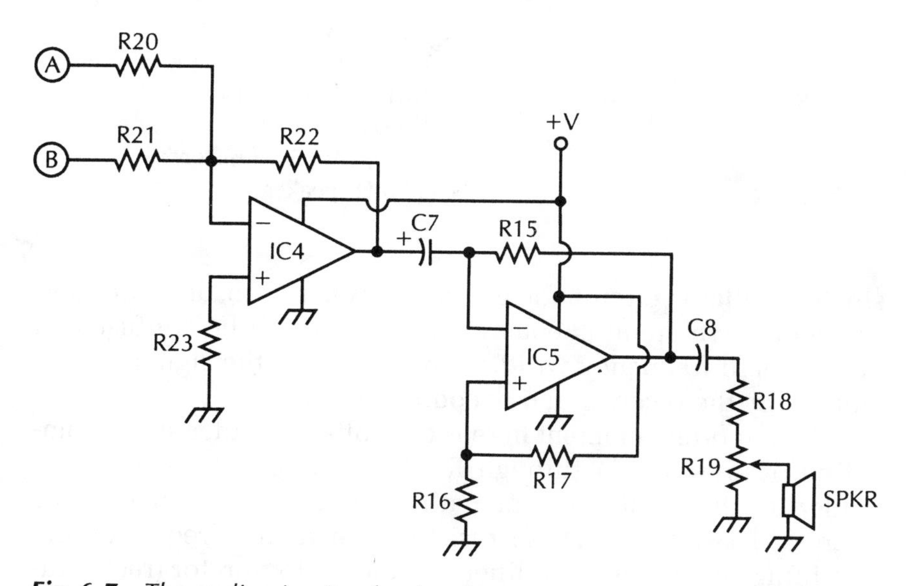

Fig. 6-7 The audio circuitry for the alternate DTMF encoder project.

When a key is depressed, this voltage appears on the appropriate row and column output lines.

IC1 generates the lower row frequencies. Because only one low frequency should ever be used at a time, you only need a

Table 6-3 Parts list for the DTMF encoder project of Figs. 6-5, 6-6, and 6-7.

IC1–IC3	7555 or 555 timer
IC4, IC5	Op amp (748 or similar)
D1	Diode (1N4003 or similar)
KB1	12-key telephone keypad
K1	Relay to suit load
C1	0.0047-μF capacitor
C2, C4, C6	0.01-μF capacitor
C3, C8	0.047-μF capacitor
C5, C7	25-μF, 35-V electrolytic capacitor
R1, R4	4.2-kΩ, 1/4-W, 5% resistor
R2, R10	3.9-kΩ, 1/4-W, 5% resistor
R3	3.3-kΩ, 1/4-W, 5% resistor
R5	10-kΩ trimpot
R6, R16, R17, R20, R21, R22	10-kΩ, 1/4-W, 5% resistor
R7, R13, R15	100-kΩ, 1/4-W, 5% resistor
R8	2.4-kΩ, 1/4-W, 5% resistor
R9	2.2-kΩ, 1/4-W, 5% resistor
R11	5-kΩ trimpot
R12	4.7-kΩ, 1/4-W, 5% resistor
R14	220-kΩ, 1/4-W, 5% resistor
R18	1-kΩ, 1/4-W, 5% resistor
R19	10-kΩ potentiometer
R23	33-kΩ, 1/4-W, 5% resistor
SPKR	Small 8-Ω speaker

single oscillator circuit for all four rows. If two or more keys are held down simultaneously, the system will be confused. A nonstandard frequency will be generated, and the signal will be ignored by the receiver and decoding circuitry.

An important element here is the voltage divider string comprised of resistors R1 through R4. The number of resistors included in the actual current path depends on which row the depressed key is in. These resistors are in the frequency determining network of a modified astable multivibrator (rectangular-wave generator) built around IC1. Trimpot R5 is used to fine-tune the overall frequency range.

In some applications calling for maximum precision, the values of resistors R1 through R4 might become critical. You might want to use precision (low-tolerance) resistors. For maximum accuracy, use four trimpots instead of the four fixed resistors shown in the schematic, and calibrate for the exact correct frequencies for each row. Start with the bottommost row (*, 0, #) and adjust trimpot R4. Then move up one row (7, 8, 9) and adjust trimpot R3. Repeat the process for the next row up (4, 5, 6), adjusting trimpot R2. Finally, adjust trimpot R1 to calibrate the frequency for the uppermost row (1, 2, 3). You must calibrate the rows in this order, otherwise the calibration for one row might be disrupted when you calibrate another row.

Capacitor C2 is included to ensure the stability of the timer. It might not be needed in all cases, but it is cheap insurance against possible stability problems. The exact value of this capacitor is not critical.

IC2 and its associated components are wired as a similar modified astable multivibrator (rectangular-wave generator) circuit. This section generates the higher column frequencies. Everything is similar here, except for the different component values used to generate the higher frequencies. Also, there are only three resistors in the voltage divider string (R8 through R10) instead of four, because you are only using three column frequencies.

Once again, because only one high frequency should ever be used at a time, you only need a single oscillator circuit for all four rows. If two or more keys are held down simultaneously, the system will be confused. Some nonstandard frequency will be generated, and the signal will simply be ignored by the receiving and decoding circuitry.

No matter which key is pressed, one high frequency and one low frequency is generated by the circuit. As long as only one key is held down at a time, these two frequencies will always be proper values for the DTMF system.

As with the low-frequency generator section of the circuit, the values of resistors R8 through R10 might become critical in some precision applications. You might want to use precision (low-tolerance) resistors here as well. For maximum accuracy, use three trimpots instead of the three fixed resistors shown in the schematic, and calibrate for the exact correct frequencies for each column. Start with the right-most column (3, 6, 9, #) and adjust trimpot R10. Then move to the center column (2, 5, 8, 0) and adjust trimpot R9. Finally, adjust trimpot R8 to calibrate the frequency for the left-most column (1, 4, 7, *). You must calibrate the columns in this order, otherwise the calibration for one column might be disrupted when you calibrate another column.

Capacitor C4 is included to ensure the stability of IC2, just like capacitor C2 with IC1. Once again, the exact value of this capacitor is not particularly critical.

Figure 6-6 shows the circuitry for a controlled relay function. Whenever a key is pressed, the relay is activated for a time period determined by the values of resistor R14 and capacitor C5. This is a basic monostable multivibrator (timer) circuit. The timing formula is

$$T = 1.1RC$$

where R is the value of resistor R14 in ohms and C is the value of capacitor C5 in farads. (Be sure to use the correct value units.) Using the component values suggested in the parts list, each time IC3 is triggered, the relay will be activated for a period equal to

$$\begin{aligned} T &= 1.1 \times 220{,}000 \times 0.000\,025 \\ &= 6.05 \text{ seconds} \end{aligned}$$

Experiment with alternate values for these two components to come up with timing periods suitable to your particular application.

Capacitor C6 is another stability capacitor, as with IC1 and IC2. Diode D1 protects the relay's coil from possible damage from back-EMF during switching. The relay itself should be selected to handle the desired load.

Of course, if your application does not need any form of relay control, you can simply omit the subcircuit shown in Fig. 6-6. In this case, eliminate the following components from the parts list: IC3, K1, D1, R14, C5, and C6.

The circuitry shown in Fig. 6-7 mixes the high- and low-frequency signals together and produces the actual audio output. Nothing is particularly critical in this part of the circuit. You can substitute an entirely different audio mixer circuit and/or audio amplifier circuit if you prefer.

IC4 and IC5 are standard op amps. Almost any type can be used. There isn't very much to be gained by using high-grade, low-noise devices in this application. I used a couple of 748 op amp ICs because I happened to have them handy. If you prefer, you can use a dual op amp chip in place of two separate units.

Resistors R20, R21, and R22 must have the same values to mix the two frequency components (high and low) at the same levels. IC4 acts as the mixer. It is nothing more than a simple inverting summing amplifier. IC5 is used as a simple audio amplifier. Again, the inverting input is used, and it really doesn't make much difference, except it is a little more convenient.

Potentiometer R19 serves as a volume control. If you don't need manual volume control in your application, you can simply eliminate R19. You might want to increase the value of resistor R18 for a lower fixed volume, if appropriate.

Capacitor C8 blocks any dc component that might appear in the output signal (dc offset voltages in inexpensive op amp devices make this a likely condition).

DTMF decoder

If you want to use a DTMF generator for remote control or non-telephone-related communications, you will need some way to decode the dual-frequency tones into the appropriate control signals. This is done with a DTMF decoder.

I will not show a complete schematic diagram for this project. If you really want a (technically) complete schematic diagram, you can make one up with a photocopy machine. This chapter includes all the information you need to build the DTMF decoder project. This particular project is made up of a relatively simple circuit that is repeated several times. It would be redundant and confusing to draw a complete schematic with so much repetition, so the basic circuit is shown once in Fig. 6-8. Just build as many copies of this circuit as necessary. (I'll get to that in a moment.)

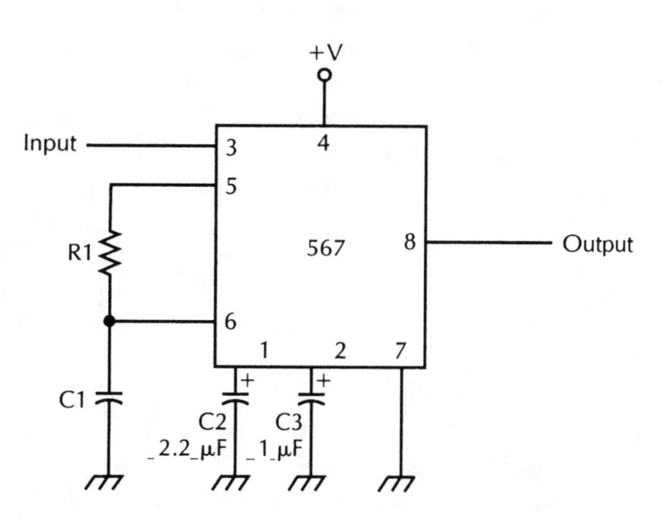

Fig. 6-8 *The basic tone decoder circuit used in the DTMF decoder project.*

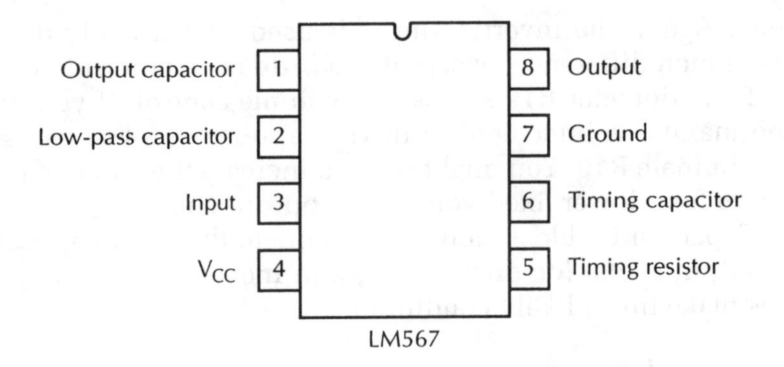

Fig. 6-9 *The pin-out diagram for the LM567 tone decoder IC.*

The basic subcircuit that is repeatedly used in this project is a simple tone decoder circuit built around the LM567 tone decoder IC. This chip might be available from other manufacturers with a prefix other than LM. If the main body of the identifying number is 567, and the IC in question is marketed as a tone decoder, it should be functionally the same device.

The pin-out diagram for the LM567 tone decoder IC is shown in Fig. 6-9. The internal circuitry of this chip is basically a phase-locked loop (PLL). Unfortunately PLLs have a widespread reputation among electronics hobbyists (and many professional technicians) as complex, difficult-to-understand circuits. This isn't really true.

A simplified block diagram of a PLL is shown in Fig. 6-10. It is basically just a VCO with feedback from its output to its control input. The VCO's output signal is continuously compared to some external reference. If the VCO's output frequency starts to drop from its reference value for any reason, the error correction circuitry produces a control voltage to increase the VCO's output frequency. Just the opposite happens when the VCO's output frequency starts to drift too high—the error correction circuitry produces a control voltage to increase the VCO's output frequency. In other words, the VCO's output frequency is automatically self-correcting. That's all you really need to know about PLLs.

In the LM567 the PLL is used to detect the presence of a given frequency in the input signal. Ordinarily the output of this device (pin 8) is high (close to V +). When the predetermined frequency of interest is detected in the input signal, the LM567's output goes low (near ground potential, or 0 V). This output signal can be used to control either analog or digital circuitry to

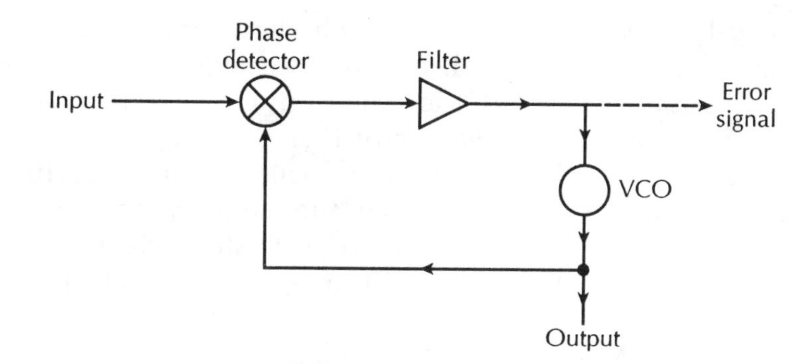

Fig. 6-10 *A simplified block diagram of a PLL.*

respond appropriately to the presence or absence of that particular frequency in the input signal.

The frequency of interest in an LM567 tone decoder circuit is usually called the center frequency, which is variously abbreviated in technical literature as "Fo" or "Fc." I will use Fc for the rest of our discussion of this device because it seems a little more logical to me.

The LM567's center frequency is determined by the values of an external timing resistor and capacitor. The formula is fairly simple:

$$Fc = 1.1/RC$$

where R is the resistance of the timing resistor in ohms, and C is the value of the timing capacitor in farads.

The functional operating range of the LM567 tone decoder is impressively wide, ranging from a low of about 0.01 Hz up to a high of about 500 kHz. In DTMF decoding you are only interested in frequencies ranging from 697 to 1477 Hz (1633 Hz if the extra fourth column is used on the keypad). This is well within the LM567's range, and you don't have to worry about going out of range in this application.

For reliable performance, the value of the timing resistor should be restricted to somewhere between 2 and 20 kΩ. Using standard resistor values, don't use anything smaller than a 2.2-kΩ resistor or larger than an 18-kΩ resistor. The value of the timing capacitor is not restricted except by the limits of the device's frequency range. In practical use, the timing capacitor's value can be anything between about 0.005 and 1100 μF. Again, for the frequency values dealt with in this particular application, you're not going to have to worry about any unusually small or unusually large capacitor values.

Now look back at Fig. 6-8, which shows the basic LM567 tone decoder circuit you will be using throughout this project. Resistor R1 and capacitor C1 are the timing components in this tone decoder circuit. The values of these components vary for each duplication of this subcircuit, depending on the particular frequency of interest for that stage of the circuitry. The values of capacitors C2 and C3 are the same for each stage: C2 is a 2.2-μF, 25-V electrolytic capacitor, and C3 is a 1-μF, 25-V electrolytic capacitor.

The basic LM567 tone decoder circuit can respond to only a single frequency. But DTMF signals are made up of two frequency components. Obviously you need two tone decoder circuits to detect and identify any particular DTMF signal. So how many tone decoder circuits do you need for a complete DTMF decoder project? Because there are twelve input keys, and each key generates a two-frequency signal, you will need twenty-four LM567s? Not quite. Remember, unique frequencies are not used for each and every key on the keypad. The keypad is divided into rows and columns. Each key in the same row has the same frequency, and each key in the same column has the same frequency. It's the combination of frequencies that uniquely identify each key's DTMF signal. Therefore, you only need one tone decoder circuit for each row frequency and one for each column frequency. The appropriate combinations can be determined with external gating circuitry. In a standard DTMF system, there are three column frequencies (1209 Hz, 1336 Hz, and 1477 Hz) and four row frequencies (697 Hz, 770 Hz, 852 Hz, and 941 Hz). This means there are only seven discrete frequencies of interest in this project, so you need seven copies of the basic LM567 tone decoder circuit shown in Fig. 6-8.

If you decide to use a four-by-four (sixteen-key) keypad in your project, you don't have to add much to the tone decoder circuit. This modification just adds one more column frequency (1633 Hz), so you only need one more LM567 tone decoder stage and the appropriate gating circuitry to combine this extra column frequency with each of the four row frequencies.

A block diagram/gating schematic diagram for the DTMF decoder is shown in Fig. 6-11. Each of the blocks marked with an identifying frequency value is a tone decoder circuit (Fig 6-8). Drawing these identical subcircuits as functional blocks greatly simplifies this diagram of the complete circuit.

A partial parts list for this project is given as Table 6-4. It only lists the components for the gating and output circuitry dis-

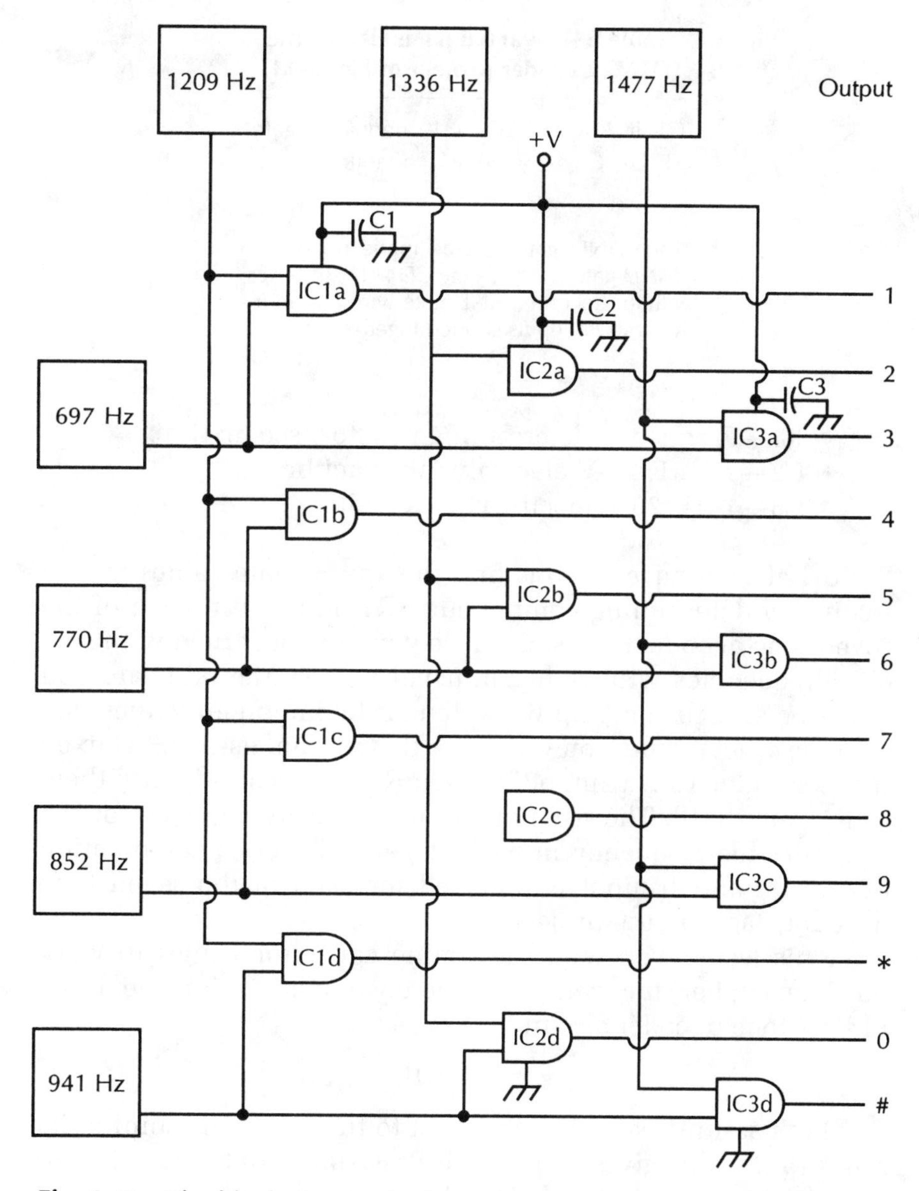

Fig. 6-11 *The block diagram/gating schematic diagram for the DTMF decoder project.*

cretely shown in Fig. 6-11. The repeated tone decoder stages (Fig. 6-8) are not listed here to avoid confusion. Remember, you need seven of each of the following components:

- IC1—LM567 tone decoder IC
- C1—Frequency determining capacitor (see text)

Table 6-4 Partial parts list for the DTMF decoder project of Fig. 6-11.

IC1, IC2, IC3	CD4081 quad AND gate*
C1, C2, C3	0.01-μF capacitor

*Each AND gate can be made up of two NAND gates (such as the CD4011) in series, with the second NAND gate wired as an inverter (both inputs shorted together).

- R1—Frequency determining resistor (see text)
- C2—2.2-μF, 25-V electrolytic capacitor
- C3—1-μF, 25-V electrolytic capacitor

Now all you have to do is find the appropriate values for the frequency determining components (R1 and C1) in each of the seven tone decoder stages. Because you are concerned with specific frequencies with rather unusual values, the odds are you won't be able to come up with standard component values that give the exact results you want. For this reason, use a 3.3-Ω fixed resistor and a 10-Ω trimpot for each R1. (You can identify them as R1a and R1b). The trimpot can be precisely adjusted for the exact desired frequency in each stage. This series combination of resistor and trimpot is repeated for each of the seven tone decoder stages in the project.

This leaves you with the timing capacitor values to work out. Remember the center frequency equation for the basic LM567 tone decoder circuit is

$$F = 1.1/R1C1$$

Let's assume each trimpot is set to the exact midpoint of its range (5 Ω). This gives a total series resistance of 3.3 Ω + 5 Ω or 8.3 Ω. For convenience, let's adjust the value of each trimpot down a little so that the series resistance is a nice, neat 8 Ω for each stage. This will be a little more convenient in your calculations. You can plug this standardized resistance value into the LM567's frequency equation:

$$F = 1.1/(8000 \times C1)$$

Now, you can algebraically rearrange this equation to solve for the unknown capacitance value in each stage:

$$C1 = 1.1/(8000 \times F)$$

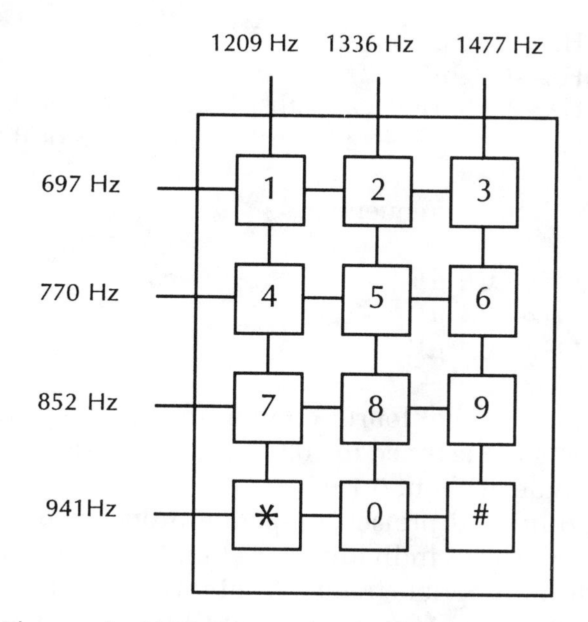

Fig. 6-12 *The standard DTMF keypad with its relevant row and column frequencies.*

Whenever you come up with a nonstandard capacitance value, you can round off to the nearest standard value and calibrate trimpot R1b accordingly. This also compensates for any errors caused by individual component tolerances. For your convenience, the standard DTMF keypad and the relevant row and column frequencies are illustrated in Fig. 6-12.

Let's start solving for the timing capacitor values with the row frequencies. The lowest row frequency is 697 Hz, so the timing capacitor (C1) in this stage should have a value of

$$
\begin{aligned}
C1 &= 1.1/(8000 \times 697) \\
&= 1.1/5{,}576{,}000 \\
&= 0.000\,000\,197 \text{ F} \\
&= 0.197\ \mu\text{F}
\end{aligned}
$$

This is not a standard capacitor value, but it is close to 0.22 μF, which is a standard capacitor value. The difference is compensated for by adjusting trimpot R1b in that tone decoder stage.

I won't step through the equations for each stage in the project. I can just summarize the rounded-off capacitor values. You'll notice that several stages use capacitors with the same rounded-off value. The difference is made up for by the calibration of the appropriate trimpot in each stage. The capacitor values for the row frequencies are

- 697 Hz—0.22 μF
- 770 Hz—0.22 μF
- 852 Hz—0.15 μF (can substitute 0.22 μF or 0.1 μF)
- 941 Hz—0.15 μF (can substitute 0.22 μF or 0.1 μF)

And, for the column frequencies:

- 1209 Hz—0.1 μF
- 1336 Hz—0.1 μF
- 1477 Hz—0.1 μF

If you use the optional fourth column in your project, the extra tone decoder stage is tuned to 1633 Hz, and uses a timing capacitor with a value of about 0.1 μF.

For the finished project to operate properly, each tone decoder stage must be individually calibrated. Feed in a known source of the appropriate frequency of interest for that particular stage, and adjust the trimpot (R1b) until the tone decoder locks onto the input signal properly. There will be some leeway on either side of each frequency. For example, a tone decoder stage tuned for 1336 Hz will also respond to 1340 Hz, and even 1350 Hz. But it shouldn't respond to either 1209 Hz or 1477 Hz, unless something is very wrong with the circuitry.

Once calibrated one and only one of the outputs should be triggered (go high) when a valid DTMF signal is fed into the circuit. All other outputs should remain low. The activated output should correspond to the key depressed on the DTMF generator/encoder's keypad. When no key is depressed (no valid DTMF signal at the input of the decoder circuit), all of the outputs should be low.

The output signals from these gates can then be fed to the appropriate control circuitry to suit your individual application. The gates should be expected to drive only very small loads directly. To drive moderate to large loads, you will need to add the appropriate relays, amplifiers, or electronic switching circuits to each output of this project.

Emergency autodialer

In most cases, it is easy enough to call for help during an emergency. Wouldn't it be nice to have an automated system that automatically calls for help if there is a fire or an intruder, whether you are home or not. That is precisely the purpose of this project.

To avoid making the circuitry too complex, there is no voice message in this project. When the circuit makes the emergency call, it doesn't tell the person who answers what is wrong. It doesn't really say anything at all. Instead, it sounds a distinctive tone to whoever answers the phone at the number that is automatically dialed. Obviously whoever answers the emergency call must already know what this tone means and what to do about it. For this reason, this project should not be set up to dial 911, the police department, or the fire department. They won't know where or what the emergency is. As far as official emergency numbers are concerned, this would be a prank, nuisance call. Not only will you not get the help you are trying to summon, you might end up facing a hefty fine or other legal consequences for misusing the official emergency communications system.

It is important enough to bear repeating. Do NOT program this emergency autodialer to call 911 or any other official emergency service (the police department, the fire department, etc.) Instead, the project should dial a friend's telephone number. Of course, the friend should be told in advance what the alarm tone means and what to do about it.

Another reason the emergency autodialer should not be programmed to call an official emergency service directly is that the circuitry has no judgement or common sense. Security systems give occasional false alarms. Even if the official emergency services understood what the automated tone call meant and sent help to the correct location, they might find they aren't really needed when they get there.

This reliability factor depends largely on the specific sensor switches you use with the project. The circuit is designed to detect and respond to a simple switch closure. It has no way of knowing what caused the switch to close. Remember, no automated security system is perfect. Occasional false alarms are to be expected. If the person at the emergency number you use with the project understands this right from the start, you are less likely to end up with problems. It is also less likely that the person at the emergency number will start reacting as if the machinery was "crying wolf."

In some cases, it might make sense to set up the system to call yourself (at the office or wherever else you usually are when you're not at home) when the emergency detection circuitry is tripped. This way you won't be dependent on someone else's good will and reliability.

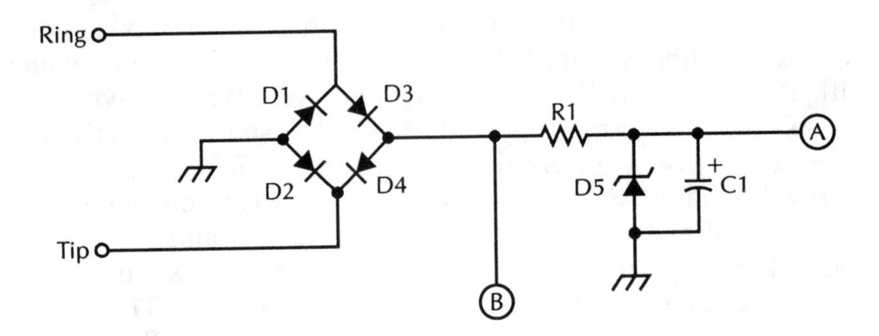

Fig. 6-13 *The input stage of the emergency autodialer project.*

Almost any normally open switching device can be used with this project to determine when an emergency condition probably exists. When an emergency condition is detected, the switch is closed, triggering the circuitry of this project. Many different intrusion detector switches are available from Radio Shack and other sources. Thermally activated switches are also available that can be used for fire detection.

You can even use a liquid-sensing switch so the project can alert you if your basement floods. As with many electronic projects, the practical applications are limited only by your imagination and your specific needs and interests.

Because of the relative complexity of this project, the schematic diagram is broken into four parts, shown in Figs. 6-13 through 6-16. You must use all four circuitry sections to have a functional project. They are split up here merely for convenience in laying out the illustrations. Attempting to cram the entire schematic diagram for this circuit onto one page would make it too small to read clearly. In constructing your emergency autodialer project, you must connect together all points marked A, B, C, and D.

The complete parts list for this project is given as Table 6-5. It is strongly recommended that you find a source for all of the required components before investing any money in this project. A few rather uncommon devices are required in this circuit. Remember that electronics is a rapidly changing field, and specialized components can suddenly become obsolete and disappear from the marketplace. In some cases, you might find a few of the discontinued parts you need at an electronics surplus house, but this can be a hit or miss proposition.

In this particular project, the most specialized component and the one that is most likely to be hard to find is IC5—the

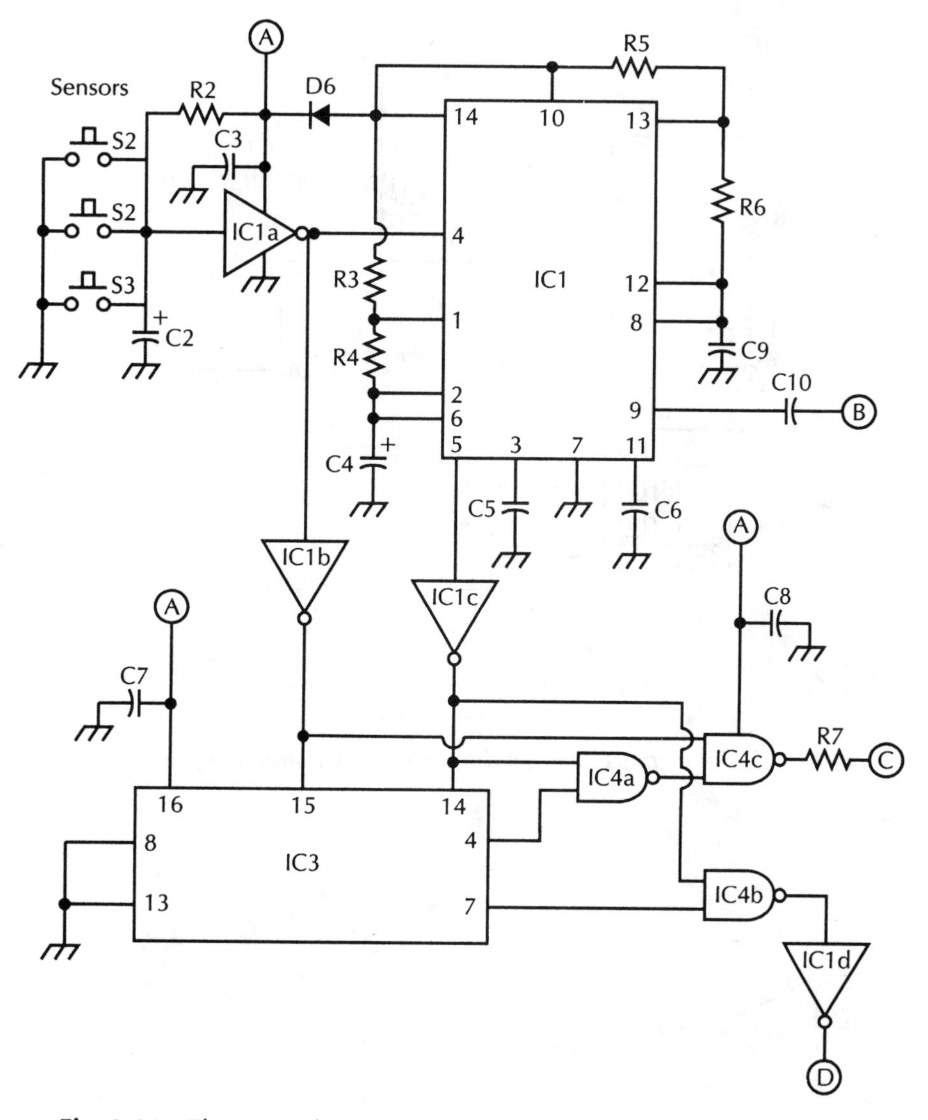

Fig. 6-14 *The sensor/control stage of the emergency autodialer project.*

MC145412P pulse/tone repertory dialer IC. This chip is designed specifically for use in autodialer circuits. It has a built-in last number redial function. That is, it can remember one telephone number (as long as continuous power is supplied) without any external memory circuitry. In this project, that is all you need. The circuit only has to remember the one number to be dialed during an emergency.

The MC145412P is manufactured by Motorola. I don't know of any direct substitute for this device, although some advanced

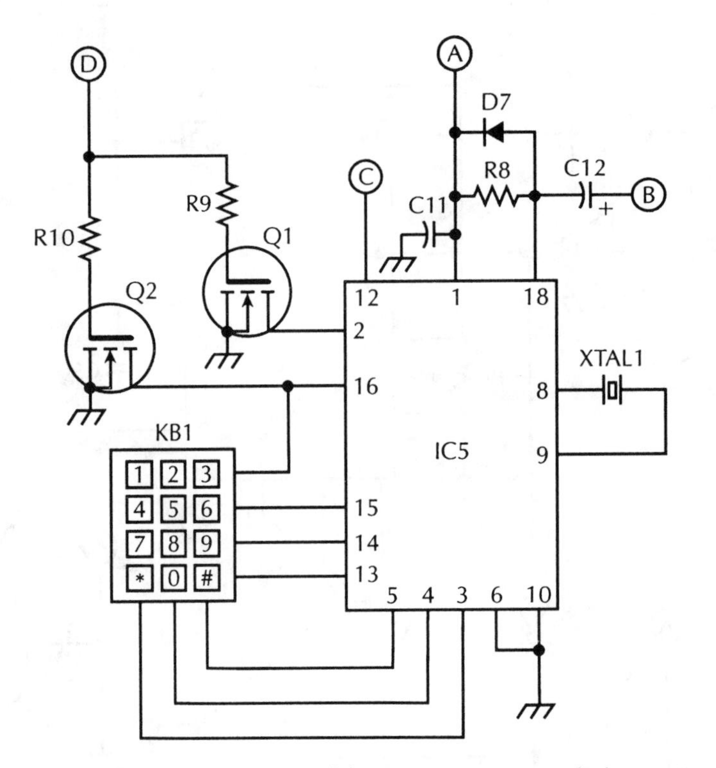

Fig. 6-15 *The memory stage of the emergency autodialer project.*

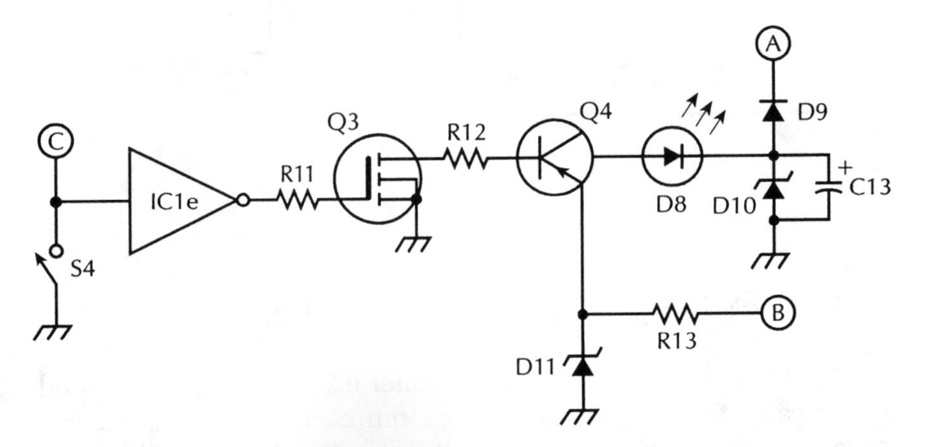

Fig. 6-16 *The final stage of the emergency autodialer project.*

experimenters might be able to modify the project's circuitry to use some other autodialer IC. Please make sure you locate this critical component before you start gathering any other parts for this project, or you might end up wasting your money.

Table 6-5 Parts list for the emergency autodialer project of Fig. 6-13 through 6-16.

IC1	CD4049 hex inverter
IC2	556 dual timer
IC3	CD4017B decade counter/divider
IC4	CD4011 quad NAND gate
IC5	MC145412P
Q1, Q2, Q3	FET (BS170 or similar)
Q4	PNP transistor (2N3906 or similar)
D1–D4, D6, D7, D9	Diode (1N4003 or similar)
D5	6.2-V zener diode
D8	LED
D10	5.1-V zener diode
D11	56-V zener diode
XTAL1	3.58-MHz crystal
C1	330-μF electrolytic capacitor
C2, C12	10-μF electrolytic capacitor
C3, C5, C6, C7, C8, C9, C11	0.01-μF capacitor
C4	47-μF electrolytic capacitor
C10	0.1-μF capacitor
C13	2200-μF electrolytic capacitor
R1	1-MΩ, 1/2-W, 5% resistor
R2, R3, R11	100-kΩ, 1/4-W, 5% resistor
R4, R7	10-kΩ, 1/4-W, 5% resistor
R5	47-kΩ, 1/4-W, 5% resistor
R6	22-kΩ, 1/4-W, 5% resistor
R8	100-Ω, 1/4-W, 5% resistor
R9, R10	470-kΩ, 1/4-W, 5% resistor
R12	15-kΩ, 1/4-W, 5% resistor
R13	220-Ω, 1/4-W, 5% resistor
S1, S2, S3, S4	Normally open SPST sensor switches (see text)

The BS170 enhancement mode field-effect transistor (FET) (Q1, Q2, and Q3) might also be somewhat difficult to find. In a pinch you can try substituting some other enhancement mode FET. The circuit should work, but I can't guaranteee it for all substitutes. Note that you must use enhancement mode FETs in this application, not the more commonly available depletion mode FETs.

Diode D11 might require a little effort to find, because many electronics suppliers dealing with hobbyists don't normally carry zener diodes above about 15 V, and a 56-V zener diode is needed here. You should be able to special order one. The exact type number isn't terribly important—it just needs to be a 56-V zener diode.

All of the other components called for in the parts list for this project are relatively common, general-purpose devices, and I doubt that you will run into difficulty locating any of them.

The 3.58-MHz crystal (XTAL1) certainly should not be hard to find, because this is the frequency used for the color subcarrier in television sets. It is probably the most commonly available crystal frequency around.

IC2 is a 556 dual timer IC. This is a commonly available device, but, if you prefer, you can substitute a pair of 555 timer chips. The 556 is the equivalent to two 555s in a single housing. This reduces the parts count and the bulk of the project. But it will work just as well with two separate 555s. For your convenience in making such a substitution, the pin-out diagram for the 555 timer IC is shown in Fig. 6-3, and the 556 dual timer IC's pin-out diagram is shown in Fig. 6-4. The pin numbers for these two devices are compared in Table 6-2.

KB1 might seem a little unusual, but you shouldn't have much trouble finding it. It is a matrix keyboard designed for use in Touch-Tone® circuits. Many devices of this type are currently available from numerous sources. In a real pinch, you can reuse the keypad from an old, broken push-button telephone, provided that the keypad is not the defective part, of course. This keyboard is nothing but a three-by-four set of matrixed switches. Pressing any one of the keys activates one of the row lines and one of the column lines, uniquely identifying the selected key. If you buy a matrix keyboard from a surplus house, you might get a four-by-four board with sixteen buttons instead of twelve. Such keyboards are used in some industrial keyboard projects. It will work just as well for this project—simply ignore the extra, fourth column, and leave its line disconnected.

In this project, the keyboard is used during the setup procedure only. It is used to enter the emergency number the circuit will dial when its sensor is triggered.

When the alarm circuit is activated by closing one of the sensor switches (S1 through S3), the MC145412P (IC5) "dials" the stored number in the form of standard DTMF tones. The telephone line responds exactly as if you were manually dialing the same number on a Touch-Tone® telephone.

The entire circuit takes its power from the telephone lines. No separate power supply is required for this project. A full-wave bridge rectifier is made up of diodes D1 through D4 to permit only ac voltages to pass through to the circuitry. Because a full-wave bridge rectifier is used, the input polarity isn't particularly critical. The project should work fine if the ring and tip connections are reversed. If by some chance, it doesn't work, it wouldn't hurt to reverse the connections to the telephone line, just in case. If the circuit functions with the connections made one way, but not the other, I strongly suspect you might have a bad diode in your bridge rectifier network.

Two output voltages are derived from the telephone line to drive the rest of the system's circuitry. The straight, unregulated, rectified input voltage is tapped off at point B. At the same time, a secondary, semiregulated voltage is derived via resistor R1, diode D5, and capacitor C1. This voltage is fed to all circuit points marked A in the schematic diagrams. It is limited by the value of the zener diode (D5), or 6.2 V. This lightly regulated voltage is used to power the ICs throughout the circuit.

A small dc voltage is continuously supplied to IC5 (MC14512P) at all times (assuming the circuit is hooked up to a live telephone line). This permits the chip to retain its memory, so it doesn't "forget" the emergency number you have stored in it. Of course, the entire project is quite useless if it can't remember the number to dial if an emergency situation is detected.

To use this emergency autodialer project, you must connect it to your telephone line. It can be connected in parallel with an existing telephone. A Y-adapter can be used. Adding this project to your telephone line should not affect ordinary telephone operation in any way. When it is not activated, the circuit draws only a tiny amount of current to maintain IC5's memory and to keep the circuit active enough to respond to an emergency situation. Only when one of the sensor switches (S1 through S3) is closed does the circuit draw any current. It is still not a high current drain. While the circuit is dialing its emergency number, it might load down the lines enough that you can't simultaneously use another telephone on the same line. But this is scarcely a practical limitation. After all, if you're home using the telephone, why can't you dial the emergency number yourself.

Close S4 to set up the circuit. Then key in the desired emergency number on the keypad (KB1). This is done exactly as if you were dialing an ordinary Touch-Tone® telephone. That is all there is to the setup procedure for this project. It might look like a very complicated circuit, especially because it is divided into four schematic diagrams, but it is actually a remarkably easy project to use.

Once the desired emergency number is entered, return switch S4 to its open position to arm the circuitry. Whenever one (or more) of the sensor switches is closed, the circuit will be activated, and it will automatically dial the stored emergency number.

What if you make a mistake in punching in the emergency number, or what if you want to change the stored number for whatever reason? No problem. Just open switch S4, then reclose it and reenter the number from the beginning. You can't change just a single digit in the stored number without erasing the entire thing and starting over.

In some cases, the memory might not be completely cleared as you reload the number. If there is any doubt, disconnect the circuit from the telephone line for a moment. This removes all power from the circuit. Without a supply voltage, the MC145412P forgets everything it ever knew. It's memory will be completely cleared.

The sensor switches should be selected to suit the specific type of emergency you want the project to watch out for. For example, normally open intrusion detector door or window switches used in burglar alarm systems are ideal. There are also sensor switches that respond to temperature, gas leaks, flooding, smoke, and almost anything else you can think of. The only restriction here is that all the sensor switches must be of the normally open type. In other words, the circuit is broken except when the sensor is tripped by the appropriate emergency condition. There is no reason why you can't combine several different types of sensor switches to guard against several different emergency conditions.

Figure 6-14 shows three sensor switches. This is an arbitrary number. All of the normally open sensor switches are in parallel. As long as they are all open (nontripped), it doesn't matter how many there are. Once one of the sensor switches is closed, completing the circuit path, it doesn't matter if a second or third (or more) of the parallel switches is also closed. As far as the circuitry of this project is concerned, there is effectively only one sensor switch. Of course, this means there is no way for the circuit to determine which sensor has been tripped.

Testing your emergency autodialer project is simple enough. After first visually double-checking all the wiring and connections (which should be done with any electronic circuit before applying power) connect the circuit to the telephone lines, and enter the emergency number as described earlier in this section.

Then arm the circuit by opening switch S4, and deliberately trip (close) one of the sensor switches. In some cases, it might be easier to add a jumper wire across capacitor C2 to simulate a closed sensor switch. If everything is working as it should, the circuit should dial the stored emergency number. When the phone at the other end is answered, the person should hear a distinctive tone, alerting them of an emergency situation in your home. (Naturally, you should notify your friend in advance whenever you test the system.)

Using the component values suggested in the parts list, the emergency tone should have a frequency of about 1000 Hz. If you would prefer to change the frequency for some reason, try experimenting with alternate values of resistor R6 and capacitor C9. This section of the circuit is simply a standard 555 astable multivibrator circuit.

Computerized dialers

In the opening to this chapter, I explained why an automatic dialer project wasn't particularly practical. But if you have a computer handy, it is not at all difficult to use it for automatic telephone dialing functions. Either pulse dialing or DTMF dialing can be supported with a computer. In either case, you will need an output port. A serial output port, or one bit of a parallel output port can be used for computerized pulse dialing. A full byte (eight bits) parallel output port is required for computerized DTMF dialing.

Computerized pulse dialing

Pulse dialing is very easy to simulate using a computer. Remember, a dialing pulse is just an on-off or high-low signal, just like the output of a digital computer. You only need one such high-low signal, so only a single output bit is needed from the computer's output port. The other bits in the output port can be applied to other uses, or simply left idle.

The key to this application is in the software. The computer must be programmed to switch the output bit between low and high in the appropriate pattern. The output signal is then fed into the telephone lines. The equipment at the telephone company's central switching office can't tell the difference between these computer pulses and the pulses from an ordinary rotary telephone dial.

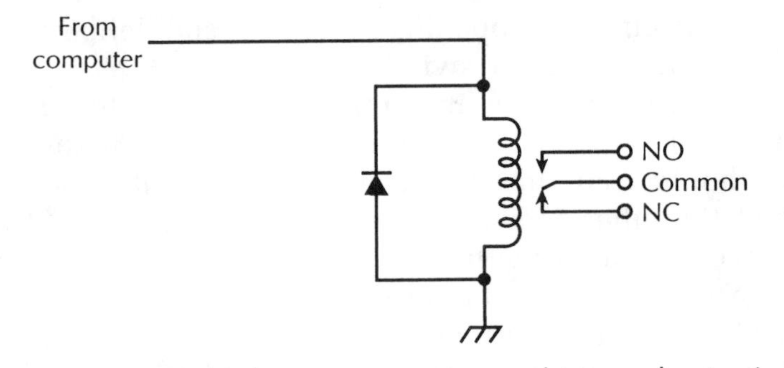

Fig. 6-17 *An output bit from a computer can drive a relay to simulate a standard dialing pulse.*

In most cases, you won't be able to use the computer's output signal directly. It probably won't be able to supply sufficient current, and the high voltage isn't the proper value. Typically a computer's output signal uses 5 V for high, but the telephone system uses about 48 V.

Fortunately there are many simple solutions to such problems. One of the easiest is to use the computer's output bit to drive a 5-V relay, as illustrated in Fig. 6-17. You can substitute a transistor switching circuit if you prefer. In this application, however, I feel the clicking of the electromechanical relay is a minor advantage. It lets you know that the circuitry is functioning properly. You can actually hear the dial pulses.

Notice that the normally closed contacts of the relay are being used to make or break the circuit with the incoming telephone lines. Resistor R1 is a current-limiting resistor. You might need to experiment with the value of this resistor to achieve the best possible results. When the computer puts out a high signal, the relay is activated, breaking the normally closed connection, just like an ordinary pulse-dialing device.

If you prefer to activate the relay with a low pulse (which might make the programming a little simpler), just use the relay's normally open contacts instead of the normally closed contacts.

Switch S1 is vitally important in this project. Without this switch, there would be no way to take the computerized dialer off-line (short of physically disconnecting it). Without this switch, the telephone company's central switching office would think that your telephone was always off the hook. Open the switch to break the connection so the system appears to be on the hook. If you don't want a manual switch for some reason, you can easily automate it. Just use another of the computer's output bits to control a second (DPDT) relay that serves the function of this switch. The computer can then automatically control when its dialer is on- or off-line. Of course, the programming must account for this modification, by holding the appropriate output high whenever the dialer is being used, and holding it low at all other times.

The software for this project must be written to suit the particular computer you are using. It shouldn't be at all difficult to write a suitable program. All the program has to do is keep the designated output bit low except during dialing pulses, when it goes high (activating the relay and simulating the broken circuit of a standard mechanical rotary dial). The pulses count up to the desired digit value. For example, to dial 935, the computer puts out a string of nine pulses, then a pause, then three pulses, another pause, and finally five pulses. Remember that in the telephone system the digit 0 counts as ten. It is represented by ten dial pulses.

The timing of these output pulses is somewhat critical. For each complete pulse cycle, the output should go high for approximately 60 ms, and low for about 40 ms. It might take some experimentation with delay loops in the program to get the timing right. Fortunately most telephone systems aren't too terribly fussy about the pulse timing, as long as it is reasonably close.

The program must also include some sort of delay loop to insert a pause between digits. This is the only way the telephone company's central switching office can tell that one digit has ended and a new digit has begun. The between digit pause can be up to several seconds long, as humans manually dialing a number will often pause to remember the next digit. You're more concerned with the minimum pause length. This will vary somewhat, depending on the telephone system in your area. You should be okay if the between-digit pulse is at least 250 ms (0.25 second) to 500 ms (0.5 second) long. When in doubt, it is probably best to err on the side of too long a pulse.

A full local telephone number is comprised of seven digits. How long will it take the computer to dial a standard seven-digit number? It's impossible to say, because of two variables—the length of the between-digit pauses and the specific digits that must be dialed. Each dialing pulse lasts the same length of time—about 100 ms—but the number of dialing pulses varies with each digit. To dial a 1 requires just a single pulse, so it takes 100 ms. A 5, however, requires five dialing pulses, so it takes 500 ms to transmit this digit.

You can calculate the maximum time it will take to dial a seven-digit telephone number if you know the length of the between-digit pauses. Let's assume your computer program inserts a 400-ms pause between adjacent digits. The longest available digit is 0, comprised of ten dialing pulses, so the number 000-0000 will take longer to dial than any other seven-digit telephone number. Notice that this is not a actual telephone number in any area, so all real telephone numbers will take less time to dial than the calculated maximum.

You have seven groups of ten pulses each, separated by six between-digit pulses. The dash (-) between the third and fourth digits is included in telephone numbers for human convenience only. It makes it easier to read and remember the number. The telephone system does not use the dash, nor is a longer pause required at this point in dialing the number.

Each 0 digit, being made up of ten 100-ms pulse cycles lasts 1000 ms, or 1 second. You have seven 1-second digits and six 0.4-second pauses, so the total dialing time for this maximum number is

$$\begin{aligned} T &= (7 \times 1) + (6 \times 0.4) \\ &= 7 + 2.4 \\ &= 9.4 \text{ seconds} \end{aligned}$$

It will never take this system longer than 9.4 seconds to dial any seven-digit telephone number. Most practical numbers will take somewhere between 6.5 and 8.5 seconds to dial.

These are pretty long times by computer standards, but they're actually faster than you could dial the number yourself. And the computer can remember as many telephone numbers as you like. You just have to tell the computer's program who you want to call, and it automatically looks up and dials the number for you.

Computerized DTMF dialing

A computer can also be used to dial a telephone number using DTMF tones. In this case, you would use one of the DTMF generator/encoder circuits presented earlier in this chapter, replacing the manual keypad with the output bits from the computer's parallel output port. Each bit in the parallel port controls one tone frequency. There are seven tones in the standard DTMF system, so you need seven output bits. The standard parallel output on most computers is eight bits wide. Let's identify the eight bits of the output byte with the first eight letters of the alphabet: ABCD EFGH.

I will leave the most significant bit (MSB) (bit A) as a permanent low value. The next three bits will be used for the three high-frequency column tones:

- B—1477 Hz (keys 3, 6, 8, and #)
- C—1336 Hz (keys 2, 5, 6, and 9)
- D—1209 Hz (keys 1, 4, 7, and *)

Similarly the four least significant bits (E, F, G, and H) are used to control the four low frequency row tones:

- E—941 Hz (keys *, 0, and #)
- F—852 Hz (keys 7, 8, and 9)
- G—770 Hz (keys 4, 5, and 6)
- H—697 Hz (keys 1, 2, and 3)

Making a given bit high (logic 1) causes that particular tone frequency to be generated. A low (logic 0) bit inhibits (turns off) the controlled tone frequency.

Theoretically you have 128 possible combinations ranging from 0000 0000 to 0111 1111 (remember bit A is always set to 0). But most of these combinations are not valid for DTMF telephone systems. The tone combinations are restricted by specific, limiting rules. Two and only two frequencies are sounded at any given time. Moreover, one of those tone frequencies must be from the high column group (bits B, C, or D), and the other can only be from the low row group (bits E, F, G, and H). For example, bits C and D should never be made high simultaneously. This is not a valid DTMF code.

The only valid output combinations for this system are as follows:

- 0000 0000—No key
- 0001 0001—1
- 0010 0001—2
- 0100 0001—3
- 0001 0010—4
- 0010 0010—5
- 0100 0010—6
- 0001 0100—7

- 0010 0100—8
- 0100 0100—9
- 0001 1000—*
- 0010 1000—0
- 0100 1000—#

All other combinations of output bits are invalid and should not be permitted by the computer's controlling program.

While there is a definite pattern to these valid bit codes, the user should not be required to keep track of them when loading telephone numbers into the software for storage and later reuse. Fortunately the computer will willingly do the work for you. When the user hits the 3 key on the computer's keyboard, the program automatically converts it to the appropriate output bit code (0100 0001).

To make writing the program a little easier, you can use the decimal equivalents for each of the acceptable binary numbers that you want to appear at the computer's parallel output port. The computer will convert the decimal values into the appropriate binary values, as shown in Table 6-6.

Remember, the computer should be putting out a value of 0 (0000 0000) when no tone is to be sent. There should be a brief pause between digits, but it can be very short. Each tone pair (digit) must be transmitted in at least 50 ms. A 50-ms pause between adjacent digits is also a good idea, although you might be able to get away with a slightly shorter pause.

Notice that with DTMF dialing, all digits have the same length, regardless of their value, unlike the digits in the pulse-dialing method covered earlier. Therefore, all standard seven-digit telephone numbers have the same transmission time.

To include a little more of a safety margin, let's assume each pair of digit tones are sent for 60 ms, and the between-digit pauses are also 60 ms, so the program can use the same timing loop subroutine for both functions. Bear in mind that this 60-ms value was selected somewhat arbitrarily. You can use a different time value, provided it is long enough to be reliably recognized by the telephone system. There is no reason to make the between-digit pause the same length as the digit tones, although it seems logical and convenient to do so.

In the example system (with 60-ms digits and pauses), dialing a standard local seven-digit telephone number (with six between-digit pauses) will take

Table 6-6 Conversion of decimal to binary values.

Key	Decimal value	Binary value
1	17	0001 0001
2	33	0010 0001
3	65	0100 0001
4	18	0001 0010
5	34	0010 0010
6	66	0100 0010
7	20	0001 0100
8	36	0010 0100
9	68	0100 0100
*	24	0001 1000
0	40	0010 1000
#	72	0100 1000

$$\begin{aligned} T &= (7 \times 60) + (6 \times 60) \\ &= 420 + 360 \\ &= 780 \text{ milliseconds} \\ &= 0.78 \text{ second} \end{aligned}$$

The entire number can be dialed by the computer in a little over three-quarters of a second. Compare that with the timing for the pulse-dialing system described earlier.

The quantity of numbers that can be stored is limited only by the memory capacity of the computer you are using. If your computer has a hard drive, you are not likely to run out of memory room.

To make the system as useful and efficient as possible, you will probably want to write the program so that the desired telephone number can be summoned up by name. It's usually helpful to permit several different variations on the name. For example, if you included the number for the author of this book, you might arrange the software so this number can be called by specifying Delton T. Horn, Delton, or Horn. You might also want to permit selecting numbers by function—for example, electronics author. You might also include my number under a subheading for TAB Books. The system can be as flexible as you choose to make it. That is the big advantage that a computerized autodialer system has over any dedicated commercially manufactured autodialer.

7 Telephone control projects

The telephone was designed for the purpose of carrying on conversations over a distance. A person at one location can hear and speak to another person at another place miles away. But like so many inventions, the telephone can be put to uses it wasn't originally intended for. What else can you use a telephone for?

Begin by generalizing what it does. It carries signals from one place to another. It doesn't particularly care what those signals are, or how they are used, provided they are more or less in the middle of the audible range of ac frequencies. Signals at a distance? That sounds like some form of remote control, doesn't it?

Yes, you can use your telephone as a handy remote control device. You just need to add the proper circuitry at the receiving end. This chapter features several projects that are designed for precisely this sort of application.

Admittedly there are some serious limitations to using a telephone for remote control purposes. But it is inexpensive, and you probably already have telephone lines installed, so if it suits the purpose, why not go ahead and use it? It's certainly less trouble than installing your own remote control cables. It's also less fuss and bother than using radio transmitters and receivers, which often require licenses, delicate calibration, and precisely mounted antennas.

The remote control projects in this chapter won't work for all remote control applications. But it is unquestionably worth experimenting with remote control functions via the telephone lines. As presented here, these telephone remote control projects

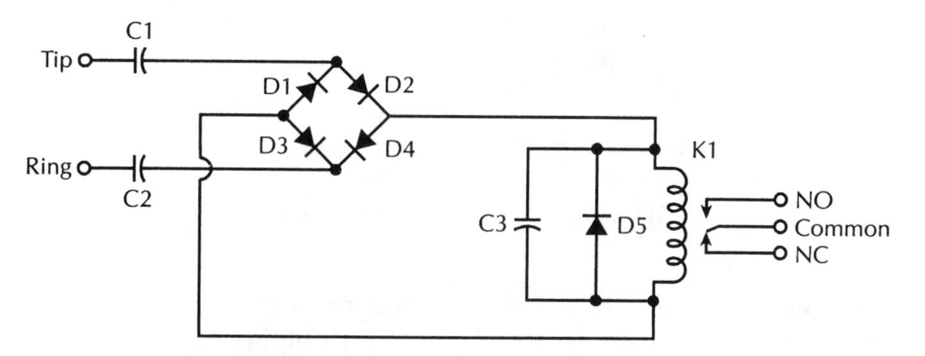

Fig. 7-1 *This circuit can turn almost any electrically powered device on or off whenever the telephone rings.*

are fairly crude and are as simple as possible. The basic concepts being dealt with here can be readily expanded into more sophisticated systems and customized to your individual requirements.

There are two basic approaches to using a telephone as a remote control device. The easiest (and crudest) approach is to use the ringer signal to activate some sort of switching circuit (such as a relay). More advanced projects use audio frequency tones sent over the telephone lines like speech signals. This chapter features projects of both types, plus a couple of special-purpose telephone-related control functions.

Telephone-activated relay

You can use your telephone for crude remote control applications. The circuit shown in Fig. 7-1 can turn almost any electrically powered device on or off when the telephone rings. The parts list for this simple project is given as Table 7-1.

This project couldn't be much simpler. Capacitors C1 and C2 block the regular dc voltage used to operate the telephone. Only the ac ringer signal gets through to the rest of the circuit.

The ac ringer signal is rectified into a dc voltage by a full-wave rectifier circuit made up of diodes D1 through D4. This voltage is fed through the relay's (K1) coil, activating it. The common connection is moved from the normally closed (NC) contact to the normally open (NO) contact. Which set of contacts you use depends on your specific intended application. If you use the normally open contacts, the controlled device will be turned on each time the telephone rings. Otherwise, the device remains off. The normally closed contacts work in just the opposite way. In this case, the controlled device is on, but it is turned off every

Table 7-1 Parts list for the telephone-operated relay project of Fig. 7-1.

D1–D5	Diode (1N4003 or similar)
K1	Relay to suit load
C1, C2, C3	0.47-μF, 250-V capacitor

time the telephone rings. The specifications of the relay you use in your project depend on the requirements of the specific device you want to control.

Diode D5 and capacitor C5 protect the relay's coil from possible damage from high-voltage spikes when the contacts are switched.

As it stands, this project is very simple, but with its simplicity comes some significant limitations. You probably won't want to use this project as it is presented here. But it is a great starting point for a more sophisticated telephone control system. You can easily modify this project, using your imagination and creativity to apply this simple circuit to your individual needs.

The chief limitation is that the relay's contacts are activated only while the telephone is actually ringing. Remember, the ring signal is turned on and off in a regular pattern. This means the device will go on and off too. And as soon as the calling party hangs up, the system shuts down.

You can use a latching relay to lock the relay's contacts into one position or the other. Another approach is to use the relay to trigger a timer circuit controlling a second relay that actually controls the final device. When the telephone starts to ring, the controlled device remains on (or off, if normally closed contacts are used) for at least the length of the preset timer period.

Of course, there is always a problem with using a telephone for control purposes. You never know who might ring your number or when. Will that matter for the device you want to control? A logical application is to use this project to turn on a light when the telephone rings in the middle of the night.

Improved telephone-activated relay

The simple telephone-activated relay of the preceding project is functional, but crude. It is hardly suitable for many practical applications (without creative modifications). One limitation of the project is that its relay is activated only when the telephone is

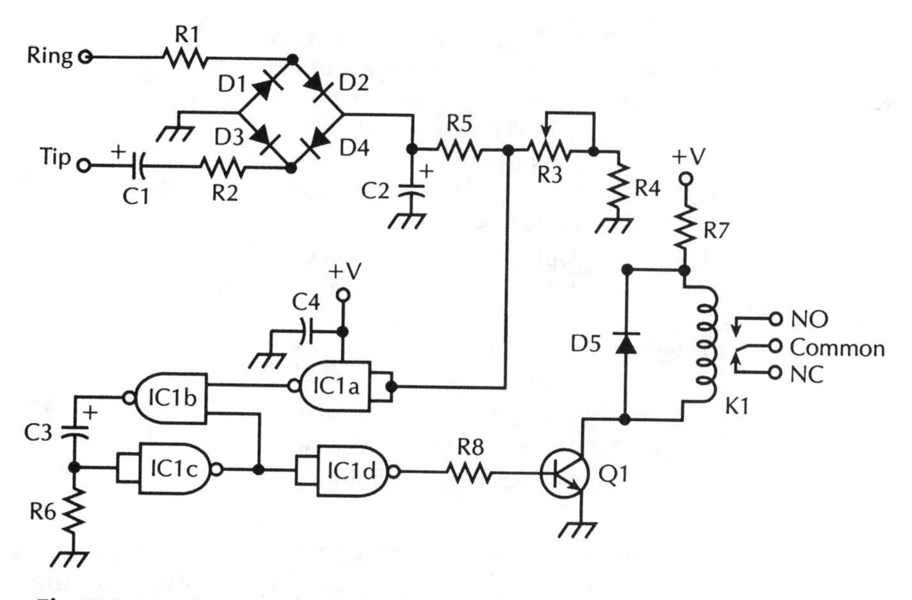

Fig. 7-2 *An improved telephone-operated relay project.*

actively ringing. But remember, the ring signal sent along the telephone lines is pulsed, not continuous. The ringer sounds for 2 seconds, followed by a 4-second pause, before sounding again for another 2 seconds. The controlled device is turned on and off in this same pattern. For example, if you use the simple telephone-activated relay project of the preceding section to turn on a lamp whenever the telephone rings, the light will flicker on and off in step with the pulsed rings.

An improved telephone-activated relay circuit is shown in Fig. 7-2. A suitable parts list for this project is given as Table 7-2. None of the component values are particular critical in this circuit, although it is a good idea to stay reasonably close to the suggested values.

This improved telephone-activated relay project uses digital gates to control the action of the relay in response to the ring signal. A slight delay (hysteresis) is added via capacitor C3. The exact delay period is determined by the value of this capacitor, along with that of resistor R6. The 100-μF capacitor and 1-MΩ resistor were selected to give an adequate delay for the standard pulsed ring signal used in the United States.

Capacitor C1 blocks any dc signal on the telephone lines. Only the ac ring signal is of interest in this circuit. It must ignore the normal operating voltages present while the telephone is in use during dialing or a conversation.

Table 7-2 Parts list for the improved telephone-activated relay project of Fig. 7-2.

IC1	CD4011 quad NAND gate
Q1	NPN transistor (2N3904 or similar)
D1–D5	Diode (1N4004 or similar)
K1	Relay to suit load
C1, C2	1-μF, 250-V electrolytic capacitor
C3	100-μF, 25-V electrolytic capacitor
C4	0.01-μF capacitor
R1, R2, R5	470-Ω, 1/2-W, 5% resistor
R3	1-MΩ trimpot
R4, R6	1-MΩ, 1/4-W, 5% resistor
R7	100-Ω, 1/2-W, 5% resistor
R8	1-kΩ, 1/4-W, 5% resistor

The ring voltage is dropped a little by current-limiting resistors R1 and R2, then it is rectified to a pseudo-dc voltage by the bridge rectifier made up of diodes D1 through D4. Capacitor C2 filters and smooths out this pseudo-dc signal a little.

Notice that a separate dc voltage source must be used to power the digital gates. The voltages on the telephone lines are too unreliable for this purpose. A large noise spike on the telephone lines (a not uncommon occurrence) can burn out the IC. This dedicated power supply can have any voltage between 6 and 15 V. I recommend using a voltage at the upper end of this range, if possible. I don't advise using a supply voltage of less than 9 V, although theoretically the circuit should work with a somewhat lower supply voltage.

Trimpot R3 and resistor R4 control the sensitivity of the circuit and drop the voltage seen at the input of the first gate. Too high a voltage here can damage the digital circuitry inside the chip. Do not reduce the size of resistor R4. Adjust trimpot R3 so the circuit responds reliably to a ring signal but ignores ordinary noise and other signals that might appear on the telephone lines. In other words, the relay should be activated reliably whenever the telephone rings, but not at any other time. If the circuit is too sensitive on a noisy line, it might respond to false signals.

A quad NAND gate is used for the control circuitry in this project. Do not substitute a TTL device. TTL gates are too demanding in their voltage requirments, both for the supply voltage and for the high input voltage. A TTL chip will almost certainly fail prematurely in an application like this. However, CMOS gates will work quite well.

Ordinarily there is no ring voltage on the telephone line, so nothing gets through the bridge rectifier. No signal is fed to the input of IC1. More precisely, the signal at this point is essentially at ground potential (0 V). This is seen by the digital gate as a logic 0, or low signal. The relay remains in its normal, nonactivated position.

When a valid ring signal is detected, a positive voltage appears at the input of IC1a. This is seen as a logic 1, or high signal, triggering the circuit and activating the relay. When the ring signal is removed the relay remains activated for a few seconds, thanks to the time delay effect of capacitor C3 and resistor R6. If another ring signal is detected before this delay period times out, the timer sets back to zero, beginning the delay period after this new ring signal is removed. The relay remains continuously activated during the normal pauses between rings. When the ring signal stops—either you've answered your telephone or the calling party has hung up—the relay remains activated only for the brief delay period of C3 and R6. After a few seconds, the circuit times out and deactivates the relay. This automatically resets the entire system so it is ready to respond the next time the telephone rings.

Almost any NPN transistor should work for Q1. This transistor only serves as a simple amplifier stage. The output of the CMOS gate (IC1) is internally limited in its current capability. It doesn't have enough "oomph" to drive any but the tiniest of relays. Transistor Q1 eliminates the possibility of unreliable relay response or possible damage to the output circuitry of the CMOS gate (IC1).

Diode D5 is included to protect the relay's coil from burning itself out with back-EMF when it switches from the active to inactive states. Select the relay to suit the desired load for your project. You might need to experiment with a somewhat different value for current-limiting resistor R7. Depending on the specific characteristics of the relay you use, you might be able to eliminate this resistor from the circuit, but do so with caution or you might end up with a damaged relay coil and/or transistor Q1. Also if you use battery power for this project, the use of a current-limiting resistor in series with the relay's coil will help extend battery life.

DTMF remote controller

The simple telephone controller projects presented so far in this chapter are functional, but extremely limited. Both of these circuits respond indiscriminately to any ring signal on the telephone lines. Because a telephone-activated controller circuit of this type responds to only one type of signal, it can only do one thing. It can turn something on or off, and that's about it.

For some applications it might be more practical to add a timer stage, to hold the control relay in an activated state for a predetermined length of time. Even if the telephone stops ringing after just one ring, the relay might remain activated for a minute or two.

This is a small improvement, but there are still some very serious limitations to this form of telephone control. The circuit responds every time the telephone rings. What if someone else calls the number? The relay will be activated. This might not be a problem, depending on your specific application.

In some cases, it might make sense to have a second telephone line installed just for control purposes. This number is only used to activate the relay, and is not given out for voice calls. But the control circuit can still be falsely triggered from time to time. Telephone solicitors often use computers to call all possible numbers at random (often creating problems for computer modems or fax lines). And then there is always the possibility of a wrong number.

Even under the best of circumstances, the control circuits dealt with so far in this chapter can only control one thing in a very limited way. Suppose you need more sophisticated or reliable remote control by telephone. A much improved telephone controller circuit is shown in Fig. 7-3. A suitable parts list for this project is given as Table 7-3.

Notice that the preliminary input stage of this circuit is the same as in the first project of this chapter (Fig. 7-1). It detects any ring signal and activates relay K1. This relay grounds pin 2 of IC1 when it is activated. Normally this pin is held high through resistor R1, unless it is forcibly shorted to ground through the relay's switch contacts.

This part of the circuit is powered by a dedicated power source of some type—not from a voltage derived from the telephone lines. This is important. The project will not work if you attempt to power it from the telephone line voltage. You can use any dc voltage from 9 to 15 V.

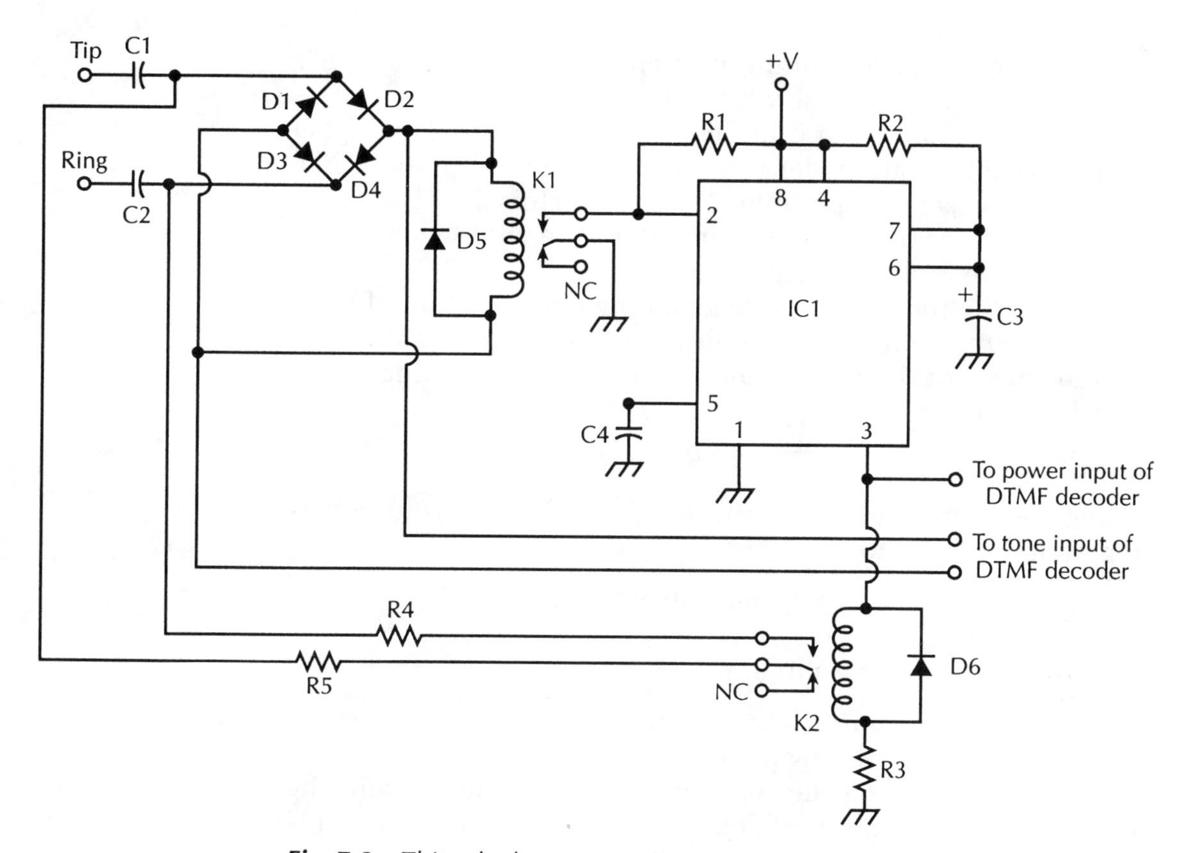

Fig. 7-3 *This telephone controller project uses DTMF signals.*

Table 7-3 Parts list for the DTMF remote controller project of Fig. 7-3.

IC1	7555 (or 555) timer
D1–D5	Diode (1N4003 or similar)
K1, K2	Relay to suit load
C1, C2	0.47-μF capacitor
C3	100-μF, 25-V electrolytic capacitor
C4	0.01-μF capacitor
R1	1-MΩ, 1/4-W, 5% resistor
R2	680-kΩ, 1/4-W, 5% resistor
R3	1-kΩ, 1/4-W, 5% resistor
R4, R5	470-kΩ, 1/2W, 5% resistor

IC1 is a familiar 555-type timer. You can use a 7555 CMOS timer if you prefer. The 555 and the 7555 are pin-for-pin compatible and require no changes in the circuitry. It is wired as a standard monostable multivibrator or one-shot timer circuit. This circuit is triggered, beginning its timing cycle, when pin 2 is brought low. This happens when relay K1 is activated.

Normally the output (pin 3) of the timer is low. This signal is used as the supply voltage for a DTMF decoder circuit, like the project presented in chapter 6. Of course, this DTMF encoder circuit receives no power and cannot operate while the timer is in its inactive (nontriggered) state.

When the timer (IC1) is triggered (by the activation of relay K1), the output (pin 3) goes high for a time period determined by the values of resistor R2 and capacitor C3, according to the standard formula:

$$T = 1.1\text{R2C3}$$

Using the component values suggested in the parts list, the timing period is equal to

$$\begin{aligned} T &= 1.1 \times 680{,}000 \times 0.000\,1 \\ &= 74.8 \text{ seconds} \\ &= 1 \text{ minute, } 14.8 \text{ seconds} \end{aligned}$$

You can substitute other values for resistor R2 and capacitor C3 to achieve different timing periods.

Besides powering the separate DTMF decoder circuit, the high output of timer IC1 also activates a second relay (K2). This closes the normally open contacts, connecting the tip and ring contacts of the telephone line through resistors R4 and R5. This simulates an off-hook telephone receiver. In effect, the circuit "answers" the ringing telephone. You might need to experiment with different values for these resistors to ensure a reliable connection that is recognized by the telephone company's central switching office.

If you call the number for the telephone line this project is connected to, you can then send DTMF control tones over the lines. You can press the buttons on a push-button telephone, or you can use a DTMF encoder/generator circuit (see chapter 6) to originate the control tones. These transmitted tones are decoded by the DTMF decoder through this circuit. You can control twelve (or sixteen) separate functions, each controlled by its own individual key.

Diodes D5 and D6 protect their respective relay coils from damage caused by back-EMF spikes. Resistor R3 is a current-limiting resistor. Again, you might have to experiment with alternate values for this component.

Capacitor C4 is included as stability insurance for the timer.

The exact value of this capacitor is not critical. Capacitors C1 and C2 block any dc component on the incoming telephone lines.

Remember, this project is not complete in itself. You need to combine it with a DTMF decoder project, and possibly a DTMF encoder/generator project. Both of these projects are discussed in chapter 6, so there is no point in repeating the circuit details here.

Stereo/TV automute

It's often hard to hear a telephone caller with the television or stereo on, especially if you tend to set the volume high. It's not a problem when you're placing the call. You can turn the TV or stereo off, or at least down, before you start dialing. But it can sometimes be a nuisance when the telephone rings while you're listening to music or watching television. It's certainly not a major problem, but it can be annoying.

The circuit shown in Fig. 7-4 provides a simple but practical solution to such problems. It automatically silences (or mutes) the television or stereo whenever you pick up the telephone's handset. This is quite a simple project and it doesn't call for very many components. The suggested parts list for this project appears as Table 7-4.

Basically this circuit is a relay with its coil connected across the tip and ring lines. When sufficient current flows through the

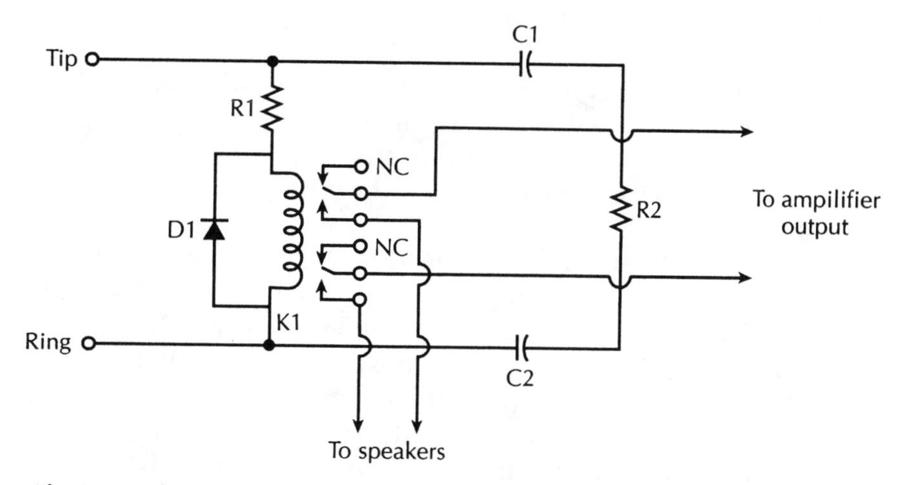

Fig. 7-4 *This circuit automatically mutes a television or stereo system when the telephone handset is picked up.*

Table 7-4 Parts list for the automatic TV/stereo mute project of Fig. 7-4.

K1	DPDT relay to suit load
D1	Diode (1N4003 or similar)
C1, C2	0.1-μF capacitor
R1	2.2-MΩ, 1/2-W, 5% resistor (see text)
R2	220-kΩ, 1/2-W, 5% resistor (see text)

relay's coil, its switch contacts are activated. The relay's normally open contacts are used in this project. The output of the amplifier is applied to the common contact, and the speaker is connected to the normally open contact. When the relay is not activated, these contacts are shorted together, and the amplifier's output is connected to the speaker in the normal way. When the relay is activated, however, this connection is broken and the speakers are disconnected from the amplifier's output. It's as simple as that.

In most audio installations, there are two wires connecting the amplifier and the speaker. You only need to break one of these wires to mute the system. It doesn't matter which one. A DPDT relay, as shown in Fig. 7-4, can be used to automatically mute a stereo system. The amplifier and speaker connections are illustrated in more detail in Fig. 7-5. If you are using the project to mute a monaural sound source (such as a television set or a radio), you won't be using both sets of connections. In this case, you can use an SPST relay in the circuit.

There is a very important caution related to this project. Some amplifier circuit designs cannot be operated safely with an infinite impedance load, which is the effect of disconnecting the speaker. In some cases, this can seriously damage the amplifier, often burning out expensive power output transistors. Obviously this is not desirable.

If your amplifier can be damaged in this way, you can easily modify the circuit by adding a dummy load resistor, as illustrated in Fig. 7-6. Notice that only a single speaker connection is shown here. The other channel is hooked up in exactly the same way. When the relay is activated, the speaker is disconnected from the amplifier's output and replaced with the dummy load resistor. The amplifier's circuitry "thinks" this resistor is a speaker, so it functions normally, without damage. The only difference is you don't hear any sound from the dummy resistor.

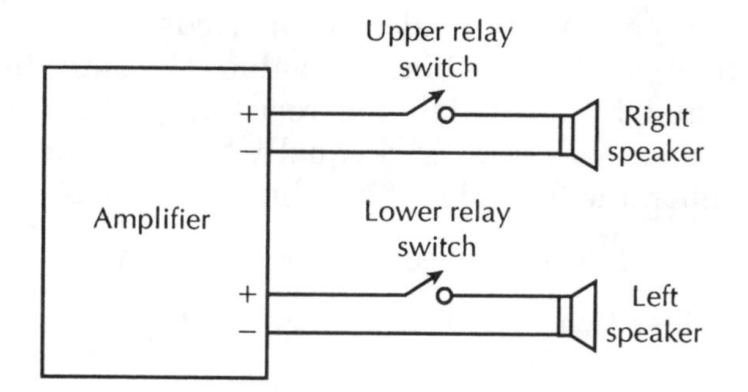

Fig. 7-5 *The proper amplifier and speaker connections for the automatic muting project of Fig. 7-4.*

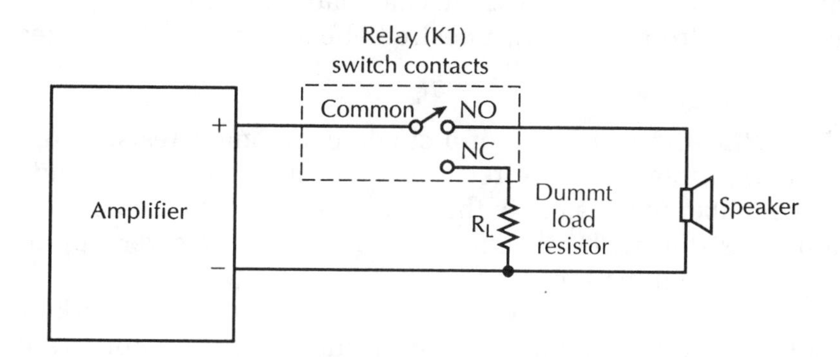

Fig. 7-6 *For some amplifiers, you need to modify the circuit of Fig. 7-4 with a dummy load resistor.*

The resistance of this component should match the nominal impedance of the speaker. This will usually (though not always) be 8 Ω. It is vitally important that the load resistor be able to handle the signal levels placed across it. It's wattage rating must be at least that of the amplifier's rated output. Preferably the resistor's wattage should be somewhat higher than the amplifer's power specification, giving some headroom just in case of the unexpected.

In larger stereo systems, the amplifier's output power can be quite large, and you are likely to have some difficulty finding a suitably hefty resistor. Try contacting someone who services industrial equipment, and see if they would be willing to sell you a couple of suitable load resistors for this type of project. Naturally these heavy-wattage resistors are considerably more expensive than the standard 1/4-W or 1/2-W units.

You can boost the effective wattage of the dummy load resistor by connecting multiple smaller resistors in parallel. The resistance for each individual resistor needs to be increased to get the correct parallel combination value. The amplifier's power level is shared by the paralleled resistors. If their values are equal, the power will be shared equally between them.

Remember, the formula for parallel resistances is

$$1/R_t = 1/R1 + 1/R2 + \ldots + 1/Rn$$

If there are just two resistances in parallel, you can substitute this formula:

$$R_t = (R1 \times R2)/(R1 + R2)$$

In this case, it is easiest if you keep the parallel resistor values equal. This makes it very easy to calculate the individual component values from the known desired effective resistance value:

$$R_x = R_t \times n$$

where R_x is the resistance of a single component resistor, R_t is the desired total effective resistance of the combination, and n is the number of resistors in the parallel combination. Each individual resistor will consume $1/n$ of the total power applied across the combination.

For example, if you want two parallel resistors to work out to a total effective value of 8 Ω, each individual resistor's value must be

$$R_x = 8\ \Omega \times 2$$
$$= 16\ \Omega$$

If each individual resistor is rated for 5 W, the combination can handle power levels up to 10 W. Similarly if you want to use five 5-W resistors in parallel to simulate a 20-W, 8-Ω speaker, each individual resistor in the parallel combination will need a value of

$$R_x = 8\ \Omega \times 5$$
$$= 40\ \Omega$$

Notice that as you increase the number of paralleled resistors, the value of each individual resistor increases.

A real speaker's impedance varies over a considerable range, depending on the constantly changing signal frequency, so you don't have to worry about getting the total effective parallel resistance exactly right. It just has to be reasonably close to the nominal value. For instance, in the last example, if you use five 42-Ω, 5-W resistors, or five 39-Ω, 5-W resistors (which would be much easier to find than 40-Ω units), the total effective parallel resistance would work out to 8.4 Ω, or 7.8 Ω (ignoring the effects of component tolerances). Either of these values is certainly close enough for this purpose. In fact, if the desired resistance is 8 Ω, you could probably get away with almost anything in the 5- to 12-Ω range with no problems at all.

You should be aware that unless the amplifier has a very, very low output power, the resistor(s) can be expected to get quite hot if the relay is held in the activated state for more than a minute or two. This is normal and does not necessarily indicate any problem, unless the resistors get too hot to touch. If this is the case, they are being overloaded, and should be replaced with higher wattage units. Because of the normal heat built up through the load resistor(s), it is important to provide adequate ventilation to give all that heat someplace to safely dissipate.

If you hook up the relay's coil across the tip and ring contacts, you'll have a problem. It will draw quite a bit of current and will "look" like an off-hook telephone to the central switching office. You won't be able to receive incoming calls, and the relay will always be activated.

To prevent this problem, current-limiting resistor R1 is added to the circuit. This resistor must have a very large value so that it draws only a minimal amount of trickle current, which will be ignored by the telephone system. When your telephone is on the hook, the relay will see the nominal +48-V signal on the telephone lines, and it will be activated. When the telephone is taken off the hook, however, the +48-V line voltage drops to a little below +10 V. This is no longer sufficient to drive the relay, which switches over to its nonactivated state.

You might need to do some experimentation to find the best value for resistor R1. If this resistor's value is too low, the project will simulate an off-hook telephone, and the normal functioning of your telephone will be disrupted. On the other hand, if this resistance is too large, the relay won't be released when the telephone is taken off the hook, and the project won't do much of anything at all.

Unfortunately, conditions vary widely for telephone systems in various areas, so I can only recommend a reasonable resistor value to get you started. If it doesn't give good results, increase or decrease the resistance with trial-and-error experimentation. The symptoms observed should tell you which way to change the resistance of R1. If the relay is always activated, the value of resistor R1 should be increased. If the relay is never activated, or the circuit performs erratically and unreliably, try decreasing the value of resistor R1.

If you find you need a very large, unwieldy value for resistor R1, you can make it up with two or more smaller (and easier to find) resistors in series. For example, a 5-MΩ resistor is pretty difficult to locate, but 2.2-MΩ and 3.3-MΩ resistors are relatively common. Using one of each in series gives a total effective resistance of 5.5-MΩ, which would undoubtedly be close enough. The required resistance value will never be precisely critical—you'll have some reasonable leeway.

For best results, use a small (low-power) relay for K1. If you are using the project to control a high-power stereo system, use the small relay to drive a larger relay or SCR circuit with its own dedicated power supply.

Any ac signal (primarily the ringer signal) on the telephone lines is shunted past the relay through capacitors C1 and C2 and resistor R2. This resistor should have a large value, but one about an order of magnitude (factor of ten) smaller than the value of resistor R1. This prevents the speakers from being turned on and off as the telephone rings. This isn't terribly important, but the effect of such "stutter-muting" can be quite annoying, and its easy enough to avoid with these three components added to the circuit.

The circuit for this project is deceptively simple, but as you can see, there is actually quite a bit to it. The basic principles of this project can be adapted to many other projects for similar telephone-control applications. The next project is built around the same core circuit that was used here.

Telephone recorder controller

Have you ever wanted to record a telephone call for personal reasons? It sounds simple enough, but it's actually much easier said than done. You can't just hold the microphone of your portable cassette recorder up to the earpiece. Well, you can try, but it is very doubtful that you'll get a decent recording that way.

You can buy special "telephone microphones" that are allegedly designed for this purpose. A cheapie microphone is encased in a suction cup body that is stuck to the telephone's handset. Frankly, all of these things that I've come across do an incredibly lousy job. Sometimes the results are even worse than the "hold the microphone close to the earpiece" method. For even marginal results, the microphone must be placed extremely close to the earpiece. But even if you place it right on the earpiece, the signal pick-up is still pathetically weak, and you'll find it very awkward trying to listen to the person on the other end of the line while you are recording the call. In my opinion, these suction-cup telephone microphones are utterly worthless.

Commercial telephone recorder controllers are available, and in most cases, work pretty well. These devices are designed to plug into a standard modular telephone jack, just as if it was an extension telephone. Because the recorder control unit is connected directly to the telephone lines, there is no need for any sort of microphone to pick up the speech signals.

Unfortunately these commercial telephone recorder controllers tend to be rather expensive. If you ever open one up, you're liable to feel you were gypped. The job can be done with very simple, inexpensive circuitry, and it makes an excellent project for an electronics hobbyist.

A circuit for a home-brew telephone recorder controller project is shown in Fig. 7-7. A suitable parts list for this project appears as Table 7-5. Nothing is terribly critical in this circuit.

This circuit has two output jacks. One is for a standard miniphone plug, which is connected to the microphone input of any portable cassette tape recorder. The voice signals are carried through this jack.

The other output jack is a smaller, microphone jack. Most standard microphones for portable cassette recorders have a built-in on-off switch, so the recorder can be started or stopped

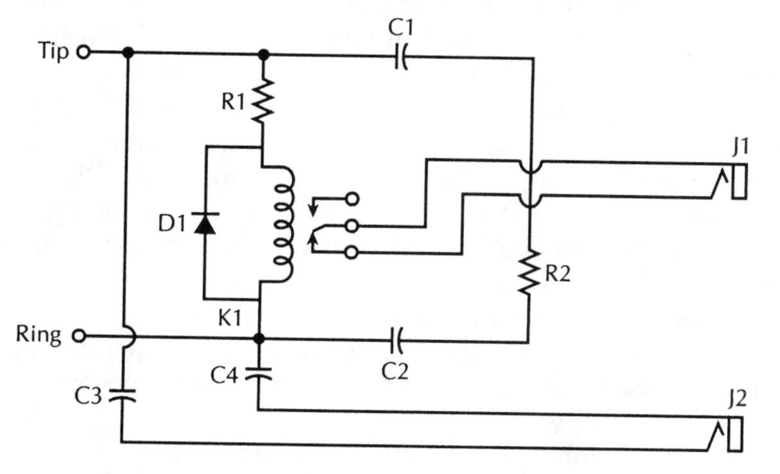

Fig. 7-7 *Telephone calls can be recorded automatically with the help of this circuit.*

Table 7-5 Parts list for the automatic telephone recorder controller project of Fig. 7-7.

K1	Small SPDT relay
D1	Diode (1N4003 or similar)
C1, C2	0.1-μF capacitor
C3, C4	0.01-μF capacitor
R1	1-MΩ, 1/2-W, 5% resistor
R2	100-kΩ, 1/2-W, 5% resistor
J1	Micro phone jack
J2	Mini phone jack

from the microphone. This is often a very handy feature. Such microphones have two plugs to connect to the tape recorder—a miniphone plug and a microphone plug. The microphone plug connects the switch on the microphone to the tape recorder's internal circuitry. When this switch is opened (in the off position), the play/record motor is electrically disabled. It has the same effect as hitting the pause button. Turning on (closing) the microphone switch permits the play/record motor to function normally, and the tape advances, continuing the recording.

In this project you are substituting the normally closed contacts of a small relay for the microphone switch, serving exactly the same purpose. Remember, as in the preceding project, that the relay is in its active state as long as the telephone is on the hook. Picking up the handset of any telephone on the line releases the relay, reverting it to its deactivated ("normal") state.

The switch contacts for this relay carry very little current, so feel free to use the smallest relay you can find. An SPDT relay is required in this project. Of course, you can use a DPDT relay (which are generally easier to find), simply by leaving the extra set of contacts disconnected. Or perhaps you might come up with a special application that will permit you to use the second set of switch contacts on the relay for some other purpose. Just remember, these two sets of contacts must always operate in perfect unison with one another because they are both part of the same relay assembly.

The microphone pause switch is under automatic control. The circuitry senses whether the telephone is off the hook or not. The cassette recorder operates only when the telephone's handset is removed from its cradle. This part of the circuit is very much like the one used in the automatic muting project presented in the preceding section of this chapter.

As in the earlier project, this circuit has a relay, with its coil connected across the tip and ring lines of the telephone jack. Ordinarily, when the telephone is on-hook, this circuit sees the nominal +48-V dc line voltage, which holds the relay in its activated condition. The normally closed contacts are held open, and the recorder's record/play function is disabled. When the telephone's handset is picked up, the line voltage drops to a little under +10 V, which is not enough to keep the relay in its activated state. The relay becomes deactivated and the normally closed contacts are shorted together, permitting the tape recorder to advance its tape and make the recording (assuming its front-panel record button is depressed).

As in the previous project, you can't just hook up the relay's coil across the tip and ring contacts, or it will draw too much current and "look" like an off-hook telephone to the central switching office. You won't be able to receive any incoming calls, and the relay will always be activated.

To prevent this type of problem, current-limiting resistor R1 is added to the circuit. This resistor must have a very large value so it draws only a minimial amount of trickle current, which is ignored by the telephone system. When your telephone is on the hook, the relay is activated by the +48-V line voltage supplied by the telephone company. When the telephone is taken off the hook, however, the +48-V line voltage drops to a much lower value. This is no longer sufficient to drive the relay, and it is released to its "normal" deactivated state.

The current-limiting resistor (R1) permits the relay to be activated without putting an excessive load on the system. Only the minimum amount of current to activate the relay is drawn. For most small relays, this current drain will be negligible and looks like ordinary leakage current drain to the telephone company's switching equipment.

You might need to do some experimentation to find the best value for resistor R1. If this resistor's value is too low, the project will simulate an off-hook telephone. The relay will always be activated, and the normal functioning of your telephone will be disrupted. On the other hand, if this resistance is too large, the relay won't be reliably triggered when the telephone is off the hook, and the project won't do much of anything at all.

Unfortunately conditions vary widely in telephone systems

in various areas, so I can only recommend a reasonable resistor value to get you started. If it doesn't give good results, increase or decrease the resistance with trial-and-error experimentation. The symptoms observed should tell you which way to change the resistance of R1. If the relay is always activated, the value of resistor R1 should be increased. If the relay is never activated, or the circuit performs erratically and unreliably, try decreasing the value of resistor R1.

If you find you need a very large, unwieldy value for resistor R1, you can make it up with two or more smaller (and easier to find) resistors in series. For example, a 5-MΩ resistor is pretty difficult to locate, but 2.2-MΩ and 3.3-MΩ resistors are relatively common. Using one of each in series gives a total effective resistance of 5.5 MΩ, which is undoubtedly close enough. The required resistance value will never be precisely critical—you'll have some reasonable leeway.

You only want the relay to respond to the drop in the dc off-hook voltage, so any ac signal (primarily the ring signal) on the telephone lines is shunted past the relay through capacitors C1 and C2 and resistor R2. This resistor should have a large value, but one about an order of magnitude (factor of ten) smaller than the value of resistor R1.

The actual audio output, connected to the tape recorder's microphone input is simply taken off across the tip and ring wires. In some cases this might confuse the telephone system, looking like an off-hook telephone. If you run into such problems, connect jack J2 across a 470- to 680-Ω resistor in series with resistor R2, instead of directly across the incoming tip and ring wires. (The nominal value of this added resistor is 600 Ω, to match the impedance of the standard cassette recorder microphone.) The problem with this method is that it somewhat reduces the signal level fed to the tape recorder.

Capacitors C3 and C4 block the dc voltage on the telephone lines and prevent it from reaching the tape recorder's microphone input, where it could conceivably cause problems, depending on the specific design of your tape recorder. With these capacitors you don't have to worry about such possibilities.

You might find that the signal level is too low for a good-quality recording. In this case you might want to add a simple audio preamplifier circuit between J2 and the microphone input of the tape recorder. In this case, you probably should use the tape recorder's auxillary input jack (usually marked "AUX") instead of the actual microphone input jack.

Don't expect a high-fidelity recording even under the best of circumstances. Remember, the frequency response of the telephone lines is intentionally restricted to just the minimum required band for understandability of speech. You can't record frequency components that have already been filtered out of the signal by the transmission process. Low audio fidelity is a built-in characteristic of telephone lines. As long as you can understand what is being said, and more or less recognize the voice of the speaker, that's probably the best you can reasonably hope for in your recording. It's certainly not a limitation of this project. The problem (if you want to call it that) is in the design of the telephone system.

You will probably find there is a very noticeable imbalance in the levels of the two halves of the conversation. Your voice is recorded much louder than the person on the other end of the line. This is because their voice signal has been dropped across the resistances and impedances of the connecting wires between their telephone and yours. You don't notice this difference in signal level when talking on the telephone, because only a small portion of your outgoing voice signal is fed back to your earpiece. The tape recorder, however, "hears" your outgoing voice signal at full strength.

Unfortunately there isn't any easy way to correct this problem. But for casual recording of a telephone conversation, this shouldn't be much of a problem.

Before leaving this project, you must consider the legal aspects of such a device. Check with your local telephone company about the applicable laws concerning the recording of telephone calls in your area. In most cases you must notify the other party that the conversation is being recorded.

Of course, if you are just recording Aunt Martha's Christmas telephone call as a family keepsake, you probably won't run into any legal problems, especially if Aunt Martha agreed to be recorded. But if you attempt to use a recording of a telephone call for anything beyond the most personal use, the legal complications can become immense. Sometimes people will record harassing telephone calls to "protect" themselves. Unfortunately they usually find that an illegally recorded telephone call cannot be used as evidence, and they might even find themselves facing a civil lawsuit or even criminal charges for making illegal recordings. Do yourself a big favor and check with a lawyer first.

Another application for this device that isn't likely to cause legal complications is to record a telephone interview, so you don't have to scribble down handwritten notes while the long-distance charges add up. If your handwriting is as bad as mine is, this can be a real help. But to protect yourself legally, first check with your local telephone company and always get the other party's permission to record the conversation. Most people don't mind being recorded during a telephone interview. But if they say no, respect their wishes. Don't try to cheat and record the conversation anyway. You could find such tricks backfiring in your face, leaving you with some expensive and messy legal problems to deal with.

❖8 Telephone amplifiers

The telephone system is designed to carry audio signals from place to place. Wherever you have audio signals, an audio amplifier seems like a good idea for an electronics project.

In a standard telephone, there is no local amplification of the received signal. There might be some boosting of the signal level at one or more of the central switching offices between the two telephones, especially during a long-distance call. But natural resistance and impedance in the connecting cables inevitably cause some degree of attenuation. The greater the distance, the greater the attenuation. Therefore, your distance from the nearest central switching office, and how far the other party (at the opposite end of the line) is from his nearest central switching office will have an effect on the reception of the voice signals.

Most modern telephones include some sort of loop length compensation circuitry to at least partially correct for ordinary distance variables, but there is rarely any local active amplification of the voice signals. In most cases, no such amplification is needed. But many people are hard of hearing and could use a little electronic help. Also people with perfect hearing might sometimes need to converse on the telephone in a noisy environment. Under such conditions, the received signal might need to be boosted a little so it can be heard clearly. That is the function of the telephone amplifier projects presented in this chapter.

In-line telephone amplifier

In most cases, it is easy enough for most of us to hear a caller's voice over the telephone. But what of the thousands of people who are hard of hearing? And all of us sometimes have to carry on a telephone conversation in a relatively noisy environment. In such cases, the normal audio levels used in the telephone system might not be sufficient. Occasionally two or more people might need to simultaneously listen in on the same telephone call. If they don't have an extension telephone, they have to crowd around a single handset, with the result that no one can hear what the caller is saying very well.

The solution to problems of this type is to add an audio amplifier to boost the audio output from the telephone, instead of relying on the built-in earpiece. This is not a difficult project. For the most part it is just a simple audio amplifier connected to the telephone line as if it was a full extension telephone. It is important to realize that this project is a receive-only device. It has no transmitter or microphone. You still need a regular telephone to talk to the party on the other end of the line in the usual way. The purpose of this project is to make the calling party easier to hear.

You don't have to worry about coming up with a high-fidelity amplifier or a wide frequency response, because the audio (speech) signals carried on the telephone lines are already intentionally limited. The important thing is for the speech to be understandable. Because the important frequency components of human speech appear in the midrange of the audible spectrum, virtually any audio amplifier circuit should do the job.

The schematic diagram for the telephone amplifier project is shown in Fig. 8-1. A suitable parts list for this project is given as Table 8-1.

The heart of the project is IC1, an LM386 audio amplifier IC. You should have no trouble finding this device, as it is quite popular, and is available from several different manufacturers. If for some reason you choose to substitute some other audio amplifier chip, only minor modifications to the circuit should be required. To make it easier to make such a substitution, the important pin numbers of the LM386 are

- Pin 6—V+
- Pin 4—ground
- Pin 3—noninverting (+) input
- Pin 2—inverting (−) input
- Pin 5—output

Pins 1, 8, and 7 are special frequency compensation pins. If you substitute another audio amplifier chip, ignore the connec-

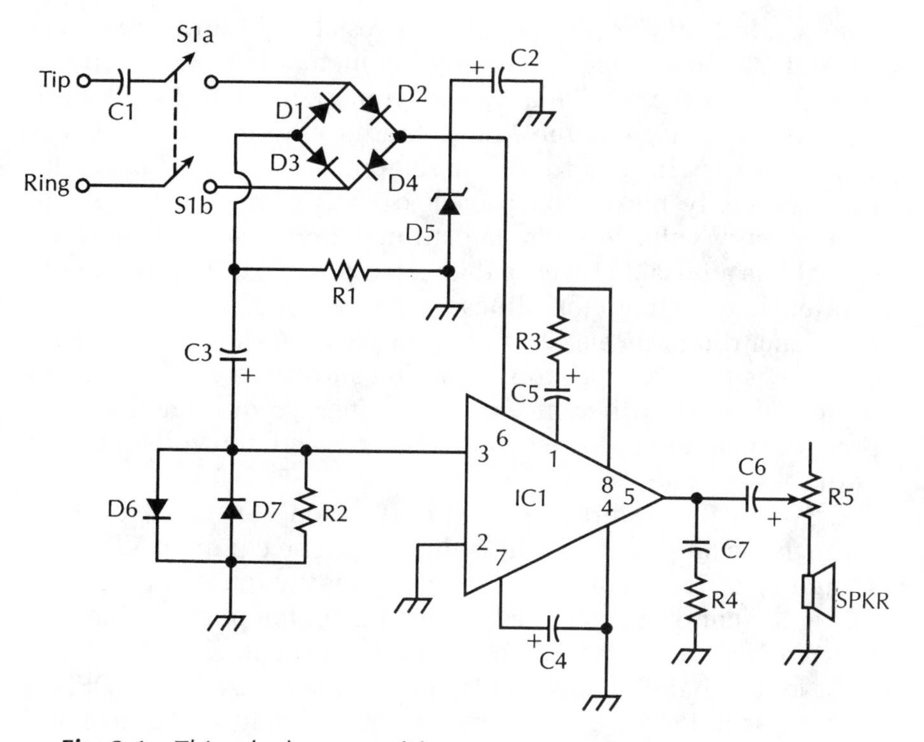

Fig. 8-1 This telephone amplifier project is designed for in-line use.

Table 8-1 Parts list for the telephone amplifier project of Fig. 8-1.

IC1	LM386 audio amplifier
D1–D4	Diode (1N4003 or similar)
D5	5.1-V zener diode
D6, D7	Diode (1N914, 1N4148, or similar)
C1	1-μF, 250-V capacitor
C2	470-μF, 250-V electrolytic capacitor
C3	22-μF, 250-V electrolytic capacitor
C4, C5	100-μF, 25-V electrolytic capacitor
C6	100-μF, 25-V electrolytic capacitor
C7	0.047-μF capacitor
R1	39-Ω, 1/2-W, 5% resistor
R2	10-kΩ, 1/4-W, 5% resistor
R3	1.2-kΩ, 1/4-W, 5% resistor
R4	10-Ω, 1/2W, 5% resistor
R5	500-Ω potentiometer (see text)
S1	DPST (or DPDT) switch
SPKR	Speaker

tions made to these pins. Eliminate resistor R3 and capacitors C4 and C5. Instead follow the recommendations given in the spec sheet for the particular device you are using in the circuit.

One advantage of the LM386 audio amplifier IC is that it consumes very little current from its power supply. This is why you can safely power the project directly from the telephone line's supply voltage. If you substitute a heftier audio amplifier, it might be advisable to use a dedicated power supply to prevent overloading the telephone lines.

Zener diode D5 acts as a very simple voltage regulator, feeding a 5-V supply voltage to IC1. (If you substitute a different amplifier IC with different supply voltage requirements, you probably won't have to change anything except the voltage rating of the zener diode.)

Switch S1 should be closed only when you want to listen through the amplifier. As far as the telephone company's central switching office is concerned, closing this switch is the same as lifting the handset from a regular extension telephone. This will prevent any incoming calls from getting through.

Notice that S1 must be a dual section (DPST or DPDT) switch. Do not try to cut corners by attempting to switch just one of the two input lines (ring and tip). It won't work and it is very likely to disrupt telephone service. A true DPST switch is relatively difficult to find, but it is easy enough to substitute a more readily available DPDT switch. Just leave the unneeded extra contacts unconnected.

Capacitor C6 blocks any dc component in the output signal from reaching the speaker, which can damage it under some conditions. Capacitor C7 and resistor R4 are a simple filter network. You might want to experiment with the values of these two components to get the best overall sound quality from your telephone amplifier. The values of capacitor C3 and resistor R2 also have some effect on the overall sound quality from this project.

Potentiometer R5 is a volume control. If you don't need such a control in your particular application, you can either substitute a low-value fixed resistor or eliminate this component altogether. In that case you connect the positive end of capacitor C6 directly to the speaker.

In some cases this circuit might not work reliably unless you omit capacitor C1 from the input circuit. You simply have to experiment to determine whether this capacitor should be used in your system or not. I recommend starting out with the input capacitor (C1) in place. If the project doesn't work, or is erratic or unreliable, try placing a short jumper wire across the capacitor (effectively eliminating it from the current path). If the project works better without the input capacitor, by all means, remove it from the circuit.

Theoretically you can modify this project into a hands-free receive/transmit telephone amplifier simply by adding a microphone to pick up the voice signals to be transmitted. Unfortunately there is a complication of potential feedback—the signal from the speaker gets fed back into the microphone and is repeatedly amplified until the circuit breaks into uncontrolled oscillation. In a hands-free receive/transmit telephone amplifier, switching circuitry is used to disable (or significantly attenuate) either the speaker or the microphone signal when the other is in use. This works because you can't really carry on a conversation if both parties are talking at once anyway. Neither can hear what the other is saying that way.

Unfortunately I have not been able to come up with a sufficiently inexpensive way to accomplish this to keep the project cost effective in comparision with commercially available units. If you need the hands-free transmit function, you'd probably be better off purchasing a manufactured unit designed for the purpose. They are not uncommon or expensive.

On the other hand, an advanced electronics experimenter might want to take up the challenge and "build a better mousetrap." How would you approach this design problem?

Earpiece telephone amplifier

For many applications, especially for the hard of hearing, the in-line telephone amplifier project presented in the preceding section might be more than is really needed. It is also not very portable. A smaller amplifier circuit might be more convenient.

This next project takes care of such complaints. The circuitry is shown in Fig. 8-2. A suitable parts list is given as Table 8-2.

Nothing is terribly critical in this circuit. If you can't find the 2N2956 transistor specified in the parts list, most other medium- to low-power PNP transistors should work okay, although the amount of amplification and the distortion level will be affected considerably by the exact specifications of the individual transistor used. In some cases you might have to modify some or all of the passive component values in the circuit to match up properly with the substitute transistor. Please save yourself some

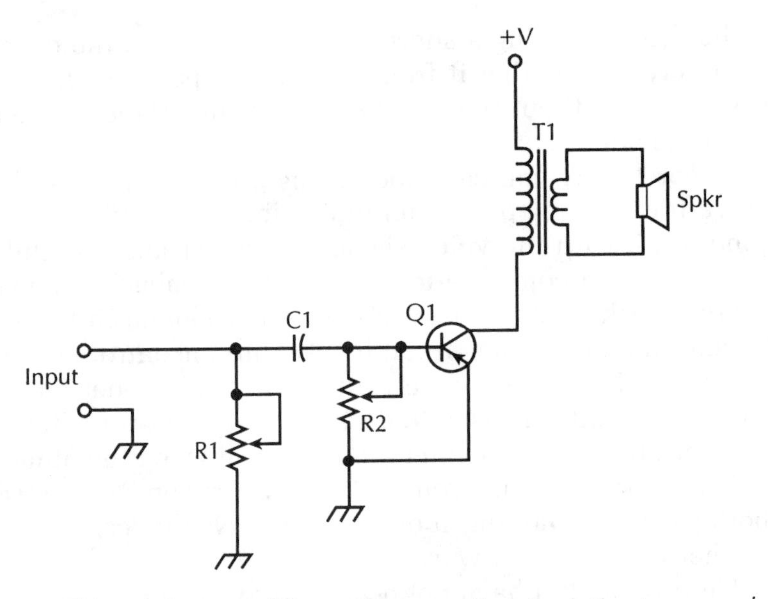

Fig. 8-2 *This earpiece amplifier project permits more intimate use and privacy, just like a regular telephone handset.*

Table 8-2 Parts list for the earpiece amplifier project of Fig. 8-2.

Q1	PNP transistor (2N2956 or similar)
T1	Impedance-matching transformer (1000:8 Ω, or to suit load)
C1	0.47-μF capacitor
R1	100-kΩ potentiometer (see text)
R2	2.5-MΩ trimpot (see text)

frustration and breadboard this circuit before attempting to build a permanent version of the project.

The output transformer (T1) is for impedance matching. The primary impedance should be about 1 kΩ (1000 Ω) to match the transistor's output impedance. The secondary impedance of the transformer should be selected to match the impedance of the speaker you are using with this project. In some (rather rare) cases you might be able to eliminate the transformer from the circuit altogether. Usually, however, this leads to excessive distortion and possibly unacceptable signal loss, defeating the entire point of the project. Experiment with the circuit on a solderless breadboard before making any permanent decision.

Two potentiometers (R1 and R2) are shown in the circuit

diagram. Potentiometer R1 serves as a volume control, and potentiometer R2 adjusts the bias on the base of the transistor.

R1 can be a manual potentiometer or a miniature screwdriver-adjusted trimpot for a "set-and-forget" volume control. If your application doesn't need an adjustable volume control, you can replace this potentiometer with a standard fixed resistor of an appropriate value.

R2 should not be a manual potentiometer. A miniature screwdriver-adjusted trimpot should be used here. In many cases, this potentiometer can be replaced with a suitably valued fixed resistor, or possibly even eliminated altogether.

The power requirements for this circuit are very low. It can work from a 1.5-V battery, but I recommend using two such cells in series for a supply voltage of 3 V. For most practical applications of this project, you will want to use the smallest batteries (in terms of physical dimensions) you can find. The AA batteries are okay, but AAA batteries or N batteries are even better (smaller), although they are generally more difficult to find. They should be available from most electronics parts suppliers. An even better choice are the tiny button batteries used in many pocket calculators and digital wristwatches. These button batteries pack a surprising amount of power into a very tiny package, which is ideal for this project. They should last a year or two under average use before they need replacement.

This amplifier circuit is designed to be as physically small as possible. It can be enclosed in a small container with a built-in microphone, and the whole unit can be affixed to the earpiece of the telephone's handset. The microphone is positioned directly over the handset's built-in receiver so it picks up the caller's voice and amplifies it through its own speaker.

You can glue the amplifier permanently in place on the handset, but this might be a little awkward. On some telephones it can be difficult to hang up the handset properly with the earpiece amplifier unit in place. A more flexible solution is to temporarily hold the amplifier in place over the telephone's earpiece with a large, wide rubber band or two. This approach also makes the project quite portable. The hard-of-hearing user can then use it on any telephone.

Some people might like the idea of the portable telephone earpiece amplifier, but would still like a permanent installation in their home telephone (or telephones). This project is inexpensive enough that it is not unreasonable to build multiple copies.

A moderately advanced electronics experimenter should have no difficulty installing this amplifier project directly inside the body of most telephones. (It is not physically possible with all models.) In the built-in version, you do not use a microphone, and the amplifier's speaker is replaced by the original built-in speaker inside the telephone.

Many telephones have a hollow handset, with screw-on caps for the mouthpiece and the earpiece. To install this earpiece amplifier project in such a telephone, unscrew the cap over the earpiece, and pull out the speaker unit. It should have two small wires attached to it. Disconnect these wires from the speaker, and solder them to the input of the amplifier circuit. Then connect the output of the amplifier to the original telephone speaker, as shown in Fig. 8-2. In constructing the amplifier you could use a very small piece of perf board cut to fit easily inside the handset, or you could use point-to-point wiring with no circuit board at all. Wrap everything with plenty of electrician's tape to make sure there is no risk of any of the component leads touching and causing a short circuit. Use tape rated for electrical work. Other types of tape might not provide sufficient electrical insulation. Fit the amplifier circuitry into the hollow handset, reposition the speaker in its original location, and put the screw-on endcap back on.

Notice that in this case, R1 and R2 must be either fixed resistors or very small, precalibrated trimpots. In some cases you might be able to mount a small potentiometer on the back of the handset to serve as a volume control (R1).

An alternate approach is to install the amplifier circuit inside the body of the telephone. You will probably have more room to work there. You will just be dealing with the other end of the speaker wires. For most simple handsets (those with nothing but a simple transmitter and receiver), there are just four wires going into the handset from the body of the telephone—two wires to the speaker (earpiece) and two to the microphone (mouthpiece). An ohmmeter or continuity tester can help you locate the correct wires. Ideally they should be color-coded, which makes things a lot more convenient.

Some modern telephones have the dialing keypad and/or other functions built into the handset instead of the body of the telephone. This means there are a lot more connecting wires leading to the handset. This type of installation might not be possible or practical in all models of modern telephones.

In an electronic telephone with active circuitry (including

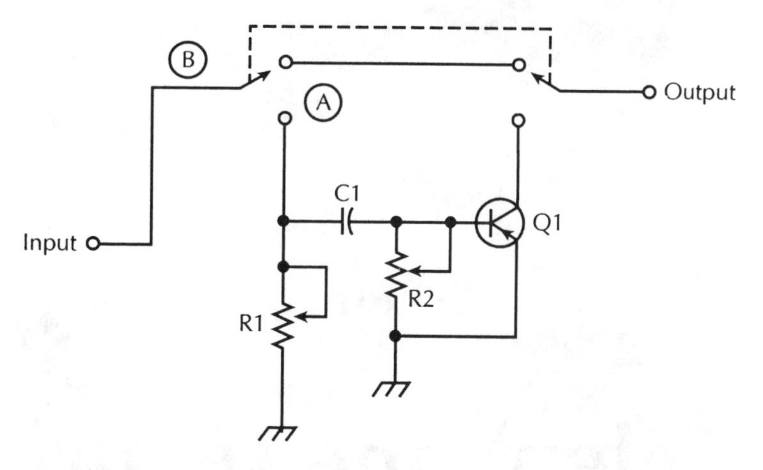

Fig. 8-3 *A bypass switch can prevent overamplification for telephone users with normal hearing.*

all Touch-Tone® telephones), there is probably a low-power voltage source you can tap into to drive the amplifier circuit instead of relying on batteries.

The one problem with such a permanent installation is that it might be annoying if other people with normal hearing use the same telephone. The received speech signals might be uncomfortably loud for a normal hearing person. If there is a manual volume control, various users can adjust the signal level back and forth to suit each individual.

Perhaps a better solution is to modify the project slightly and add a bypass switch, as illustrated in Fig. 8-3. When this switch is in position A, the amplifier circuit functions normally. If the switch is in position B, however, the voice signals are routed around the amplifier as if it didn't even exist. In this switch position, the telephone functions as if you never installed the earpiece amplifier project. All users (hearing impaired and normal hearing) get the best of both worlds with just the flick of a switch, which can be mounted in any convenient position on the handset or on the body of the telephone, depending on where you physically installed the amplifier circuit.

This is a very simple project, with just a minimum of components, but it can do a big job.

9
Telephone security

This chapter is about security in your telephone system. I'm not talking about exciting and mysterious cloak-and-dagger stuff. I assume that most of my readers don't have any important military or industrial secrets requiring the use of complex voice scramblers or bug detection systems.

The kind of security I am considering here is more mundane, but it is still very important. I am talking about the security of keeping track of your system and ensuring that it is being maintained and used in appropriate or approved ways. No system is secure unless you know what is going on in it. The various monitor circuits presented in this chapter let you know what is going on in your telephone system. Note that not all of these projects are likely to be needed by any one person, except perhaps, a confirmed paranoid. But you might find that one circuit that perfectly suits your individual needs.

Ring detector

Sometimes you might want to know whether or not someone has called you while you were out, but you don't have an answering machine. The circuit shown in Fig. 9-1 provides a simple answer to this question. The parts list for this project appears as Table 9-1. Nothing is terribly critical here, so you can substitute other component values that are reasonably close to the ones given here.

Do not use ordinary ceramic disc capacitors for C1 and C2. The metalized film capacitors in the parts list are strongly recom-

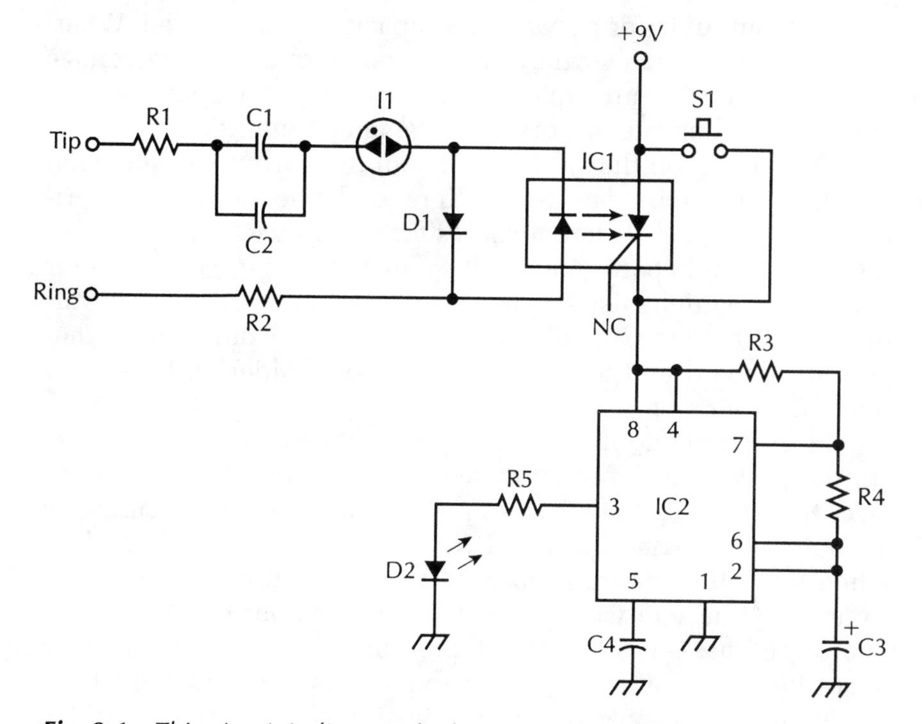

Fig. 9-1 *This circuit indicates whether or not the telephone has rung since the last time you checked.*

Table 9-1 Parts list for the ring detector project of Fig. 9-1.

IC1	Optoisolator with SCR output
IC2	7555 or 555 timer
D1	Diode (1N4003 or similar)
D2	LED
I1	NE-2 neon lamp
C1, C2	0.22-μF, 200-V metalized film capacitor
C3	10-μF, 25-V electrolytic capacitor
C4	0.01-μF capacitor
R1, R2	470-kΩ, 1/2-W, 5% resistor
R3	56-kΩ, 1/4-W, 5% resistor
R4	47-kΩ, 1/4-W, 5% resistor
R5	330-Ω, 1/4-W, 5% resistor
S1	Normally open SPST push-button switch

mended, but mylar or polystyrene capacitors can be used. Whatever capacitors you use, they must be rated for a working voltage of at least 200 V, and preferably higher, to protect against possible high-voltage noise spikes on the telephone lines.

Depending on the specifications of the particular optoisolator (IC1) you use in your project, it might be necessary to experiment with slightly different values for resistors R1 and/or R2 for the most reliable performance. In some (rather rare) cases, better performance might also be possible with somewhat adjusted values for capacitors C1 and C2. It is strongly recommended that you breadboard and test this circuit before soldering together a permanent version.

Please don't form any unrealistic expectations about this simple project. It is not an answering machine. It does not answer the ringing phone. It does not record either an incoming or an outgoing message. It gives no indication of who called or when the call came in. It doesn't even tell you how many calls came in while you were gone. This circuit does one thing and one thing only—it alerts you that your telephone rang at least once since the last time it was manually reset (via switch S1).

Perhaps the most useful application for this project is for voice mail subscribers. If this project tells you there have been no incoming calls (rings), you don't need to bother calling in on the off chance there might be some (possibly urgent) messages waiting for you.

This circuit ignores everything on the telephone line except the relatively high-voltage ring signal. All other telephone signals have too low a voltage to permit the neon lamp (I1) to conduct, so the circuit just sits there. A ring signal on the telephone line supplies sufficient voltage to fire the neon lamp, permitting it to conduct current to IC1.

When this circuit detects a ring signal, the internal LED in the optoisolator (IC1) lights up, triggering the internal SCR. Normally this SCR blocks the +9-V supply voltage from IC2, but when it is triggered, the SCR acts like a closed switch. Even when the triggering ring signal is removed, the SCR continues to conduct until it is manually reset by briefly closing and reopening switch S1. A normally open push-button switch is used here.

IC2 is a common 555 timer. For even lower current consumption, you can substitute a 7555 CMOS timer chip. The 555 and the 7555 are pin-for-pin compatible and this substitution requires no changes in the surrounding circuitry.

This timer is wired as a simple low-frequency astable multivibrator or rectangular-wave generator. As long as power is applied to IC2 (through the triggered SCR inside optoisolator IC1), the LED (D2) flashes on and off at a regular rate determined by the values of resistors R3 and R4, along with capacitor C3. Of course, you can experiment with different component values to change the flash rate. Don't use too high a flash frequency, however, or the LED will blink on and off too fast for the human eye to distinguish the individual flashes—the LED will appear to be continuously lit. A flashing LED is a much more eye-catching indicator for an application of this type. Using the component values suggested in the parts list, the flash rate for the LED is about once per second (1 Hz).

The value of resistor R5 limits the current through the LED (D2) to a safe value. Increasing the value of this resistor reduces the brightness of the LED when lit. Similarly, reducing the resistor value increases the LED's brightness somewhat. However, do not reduce the value of R5 below 100 Ω, or the LED will draw too much current and will burn out prematurely.

Capacitor C4 is included to ensure stability in the 555 timer's internal circuitry. This capacitor isn't always needed, but it won't do any harm and it is cheap insurance against possible instability problems. I believe it is a good idea to use the stability capacitor in any 555 timer circuit that does not use the voltage control pin (pin 5). The exact value of this capacitor is not critical. Experimenting with alternate capacitor values will not affect the operation of the circuit in any noticeable way.

Using this project is quite simple. It is connected to your telephone line just as if it was an extension phone, or in parallel with any existing telephone. Before you leave the house (or your office, or wherever you have the project installed), make sure the circuit is reset by briefly depressing push-button switch S1 then releasing it. The LED should be dark.

When you return, merely glance at the LED. If it is still dark, your telephone didn't ring while you were gone. If it is blinking on and off, you know that you got at least one call since you left.

This circuit is quite simple, so it is very easy to troubleshoot any problems that might crop up. If the project responds to incoming calls, but the LED is continuously lit instead of blinking, either the timer circuit is set for too high a frequency or the 555 timer chip is defective. The second possibility is highly unlikely, but I have seen such things happen. If the flash frequency is too high, simply increase the value of capacitor C3, resistor R3, or resistor R4 to slow the LED's blinks down to a visible rate.

If the LED remains blinking, even if you attempt to reset the circuit, either switch S1 is bad (you can easily check this with an ohmmeter or continuity tester) or you have a defective optoisolator (IC1).

If the project does not respond to incoming ring signals, you might have a bad LED or SCR in the optoisolator (IC1), but this is unlikely. A more likely cause is a burned out neon lamp (I1). Diode D1 could also be the culprit. In some cases the reset switch might be shorted, and this too would prevent the circuit from operating properly. Of course, make sure you have a good 9-V power source driving the timer circuit when the ring detector is operating. If the battery (or other power source) is bad, the LED can't light up.

If your ring detector's response is erratic or unreliable, try experimenting with alternate values for the resistors and capacitors in the input network (R1, R2, C1, and C2). The most likely problem is false signals when the telephone doesn't ring. Increasing any or all of these component values a little should cure the problem. If your telephone lines are subject to a lot of noise spikes, you need more input capacitance (C1 and C2). On the other hand, if your circuit sometimes ignores a proper ring signal, you need to reduce one or more of the resistor values.

You can easily check the project's response simply by hooking it to your telephone line, resetting it, and waiting for a call to come in while you are there. (It would probably make sense to have a friend call you for this purpose so you don't have to wait too long.) When your telephone rings, make sure the LED starts to flash.

Once triggered, this circuit ignores all further incoming calls until it is manually reset. You can use your telephone or any other equipment on the line while this circuit is operating.

This project draws only a very small current from the telephone lines. It shouldn't draw any current at all, except during an incoming ring signal. Even then, the circuit should draw only enough current to fire the neon lamp (I1) and the internal LED in the optoisolator (IC1). The timer (IC2) and the flashing output LED (D2) do not use power from the telephone lines, but from a separate, independent voltage source. A 9-V transistor radio battery works very well. Unless you are gone for very long periods (in which case this project is almost certainly inadequate to your needs), the battery should last a very long time. When the circuit is reset (LED D2 is dark), it draws no current from the battery at all. Even when the circuit is triggered (LED D2 is flashing), the current drain is minimal, especially if a CMOS 7555 timer is used.

No this project doesn't do very much, and its potential applications are limited. But it does a very good job for its specialized purpose at a minimum cost. It is a very simple project to build and use.

Extension dialing indicator

Have you ever wanted to know when somebody was dialing out on an extension phone on your line? For example, in a home with teenagers, you might want to limit their calls.

This project causes an LED to flicker whenever any telephone on the line is dialed. However, it should be noted that this project will function only with rotary-dial telephones. To monitor outgoing dialing with a Touch-Tone® telephone requires somewhat more sophisticated circuitry.

The simple circuit for this extension phone dialing indicator project is illustrated in Fig. 9-2. The suggested parts list for this project appears as Table 9-2. Nothing is particularly critical here, except all the components must be able to withstand the typical voltage and power levels on the telephone lines. That is, it is strongly recommended that you use at least 1/2-W resistors and a capacitor with a working voltage rating of at least 250 V (and preferably a bit higher). The diode (D1) should also have a PIV rating of at least 250 V.

When the telephone is off the hook, there is a dc voltage across the ring and tip lines. This voltage has a nominal value of 50 V. (The exact voltage might vary somewhat from area to area.)

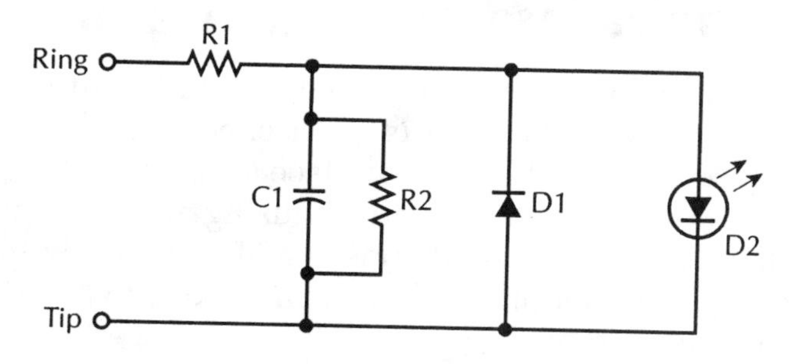

Fig. 9-2 *This project lets you know when an extension telephone on the same line is being dialed.*

Table 9-2 Parts list for the extension dialing indicator project of Fig. 9-2.

D1	Diode (1N4003 or similar)
D2	LED
C1	0.1-μF capacitor
R1	1-MΩ, 1/2-W, 5% resistor
R2	100-kΩ, 1/2-W, 5% resistor

When you dial a number, this voltage is interrupted for the appropriate number of times. For example, when you dial 6, this causes six interruptions in the off-hook voltage. Dialing 0 interrupts the off-hook voltage ten times. These interruptions in the dc voltage are detected at the telephone company, where they are decoded to make the appropriate connections via a complex system of relays and electrical switches.

When an extension telephone is taken off the hook while this project is connected to the telephone lines, the LED (D2) lights up. Each time the off-hook voltage is interrupted by a dialing pulse, the LED blinks off and back on again.

In some telephone systems, the off-hook voltage might be too low to permit this project to operate reliably. If you run into such problems, try reducing the value of resistor R1.

You can try replacing the LED (D1) with the input of an optoisolator. Then you can use the detected dialing pulses for remote control or any purpose you want. For example, you can interface this circuit with a computer that counts the dialing pulses and keeps a record of each number dialed out, along with the start and stop times of each call.

Telephone eavesdropper indicator

In a household with more than one person and multiple extension telephones, problems can often rise. Someone might pick up the handset of one telephone to place a call, only to discover that another family member is already using another extension. This is annoying and often embarrassing. If your family includes a chronic eavesdropper (many do), you are likely to be extra sensitive to such occurrences.

The circuit shown in Fig. 9-3 offers a convenient solution to help head off such problems. The suggested parts list for this

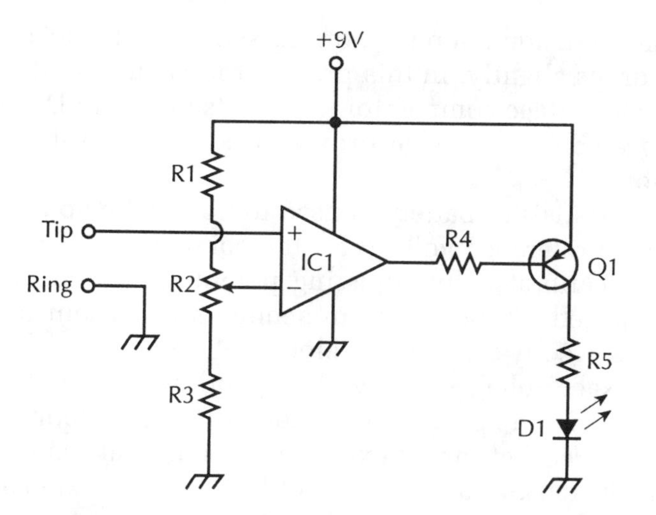

Fig. 9-3 *Telephone eavesdroppers can be detected with this circuit.*

Table 9-3 Parts list for the telephone eavesdropper indicator project of Fig. 9-3.

IC1	Op amp (748, 741, or similar)
Q1	PNP transistor (2N3906 or similar)
D1	LED
R1, R3	10-kΩ, 1/2-W, 5% resistor
R2	10-kΩ potentiometer
R4	1-kΩ, 1/4-W, 5% resistor
R5	330-Ω, 1/4-W, 5% resistor

telephone eavesdropper indicator project is given as Table 9-3. Nothing is too critical in this circuit, and many component substitutions are possible. While I have not made extensive tests, most low-power PNP transistors should work for Q1. If you are in doubt, breadboard the circuit first to make sure it works with a particular transistor before you heat up your soldering iron. With some transistors you might need to modify the value of resistor R4 for proper operation. You might also consider experimenting with different values of current-limiting resistor R5 to change the brightness of the LED.

When the telephone handset is hung up, there is about 50-V dc from the tip (green) to the ring (red) connections. The tip wire is positive with respect to the ring wire. When the telephone is in use (off the hook), this voltage drops to about 5-V dc.

If an extension telephone on the same line is picked up, the voltage drops slightly. In this project, the op amp (IC1) is wired as a simple voltage comparator that lights up the LED (D1) when it is triggered by a voltage drop caused by an off-hook extension phone.

A 9-V transistor battery is used to power the op amp and to provide the reference voltage for the comparator. The reference voltage is calibrated by adjusting potentiometer R2. A screwdriver-adjusted trimpot is recommended for this control. Once it has been correctly set, leave it alone.

The exact voltages will vary depending on your telephone company and the specific conditions on your telephone lines. Find the correct setting by experimentation. Make sure the LED is normally dark and lights up only when an extension telephone is picked up.

In some cases you might be able to eliminate the potentiometer altogether. Simply replace R1, R2, and R3 with two fixed resistors, with the connection to the inverting input of the op amp made to the junction between the two resistors.

You might want to add a switch to this project to disconnect the circuit from the telephone line when it is not in use. Otherwise the normal 50-V signal between the tip and ring wires when the telephone is on the hook will cause an imbalance in the circuit, resulting in the LED lighting up. This indication doesn't mean much of anything, but it will cause the battery to run down a lot quicker than it should. Use a double-pole switch to cut the power supply connection to the op amp and the input voltage (from the tip connection). Feeding an input voltage (especially the 50-V on-hook voltage) into an unpowered op amp can damage it.

For the security conscious, this project also detects the use of simple bugging devices that might be connected to your telephone lines, even if you have no extension phones. The bug essentially acts like an extension telephone on the line. Of course, more sophisticated bugging devices require more sophisticated detection circuits, but how many of us are likely to have our telephone lines bugged? Our telephone conversations usually aren't all that important. A simple phone tap will usually be the worst most of us are ever likely to encounter, and this project will detect a simple tap quite well.

Off-hook indicator

If you have several extension telephones (or other related equipment that electrically simulates an off-hook telephone), you can run into occassional problems. You might pick up the handset to place a call and discover the line is already in use on a different extension. For example, if you have a computer communicating by modem over the telephone lines, the simple act of picking up the handset of an extension telephone can totally disrupt the data.

Another type of problem is missed calls caused by leaving an extension telephone off the hook. Perhaps more likely is forgetting to turn off some accessory equipment connected to the telephone lines. Many of the projects in this book simulate an off-hook telephone unless they are manually turned off.

Sure these problems are more in the nature of nuisances than real disasters. But fortunately it is easy enough to avoid such annoyances. That is the purpose of this project. It is a simple circuit that turns on an LED when any telephone or other device on the line is off the hook or otherwise activated. If the LED is on when you are not using the telephone, you can easily check each telephone and device in your system. The schematic diagram for this project is shown in Fig. 9-4. A suitable parts list for this project is given as Table 9-4.

Nothing is terribly critical in this project. You should be able

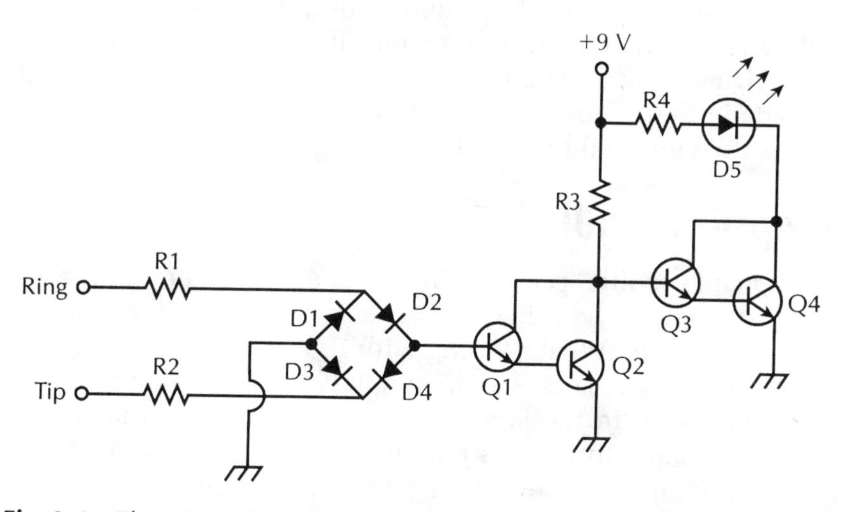

Fig. 9-4 *This circuit lets you know when an extension telephone has been left off the hook.*

Table 9-4 Parts list for the off-hook indicator project of Fig. 9-4.

Q1–Q4	NPN transistor (2N3904 or similar)
D1–D4	Diode (1N4003 or similar)
D5	LED
R1, R2, R3	10-MΩ, 1/2-W, 5% resistor
R4	1.2-kΩ, 1/4-W, 5% resistor

to find suitable substitutions for any components you have trouble locating. Almost any low-power NPN transistors can be used for Q1 and Q4. They should all be of the same type, but other than that, there aren't any real restrictions on the operating parameters of these transistors.

Basically the four transistors are grouped to operate as two Darlington pairs in a sensitive switching circuit. When the off-hook line voltage is detected at the input, the transistors are turned on, permitting current to flow through the LED (D1).

Notice that a dedicated power supply must be used with this project. It cannot be operated directly from the telephone line voltage. A simple 9-V battery does the job nicely. The current drain in this circuit is pretty low, so a battery should last quite awhile before it needs replacement. It's easy enough to check the battery (and the circuit in general). Just lift the handset of any telephone on the line. The LED should light up. When you hang up the handset (assuming nothing else on the same line is in use), the LED should go out. It's as simple as that.

Resistor R4 is a current-limiting resistor to prevent the LED from attempting to draw more current than it can safely handle. This could damage it or even burn it out.

Telephone call timer

Many people really "get into" telephone calls. They don't just say what needs to be said, but can talk on and on for hours. If they enjoy it, fine. What's wrong with it?

In some (not all) cases, there might be something wrong with it. Someone else in the family might be waiting (perhaps with great impatience) to use the telephone. Even worse, you might be tying up the line, preventing other calls (perhaps a very important call) from getting through. And many procrastinators use the telephone as a convenient excuse to keep from doing things they need to do.

In such cases it would certainly be worthwhile to limit your telephone calls to a specific length of time. This means you need to keep track of the time while you are talking on the telephone. You can try to keep an eye on the clock, but it's all too easy to get distracted and lose track. You might forget what time the call started. For that matter, how many of us remember to check the clock each time we answer a ringing telephone or place an outgoing call?

The schematic diagram for a telephone call timer circuit is shown in Fig. 9-5. A suitable parts list for this project appears as Table 9-5.

Essentially this circuit consists of just two monostable

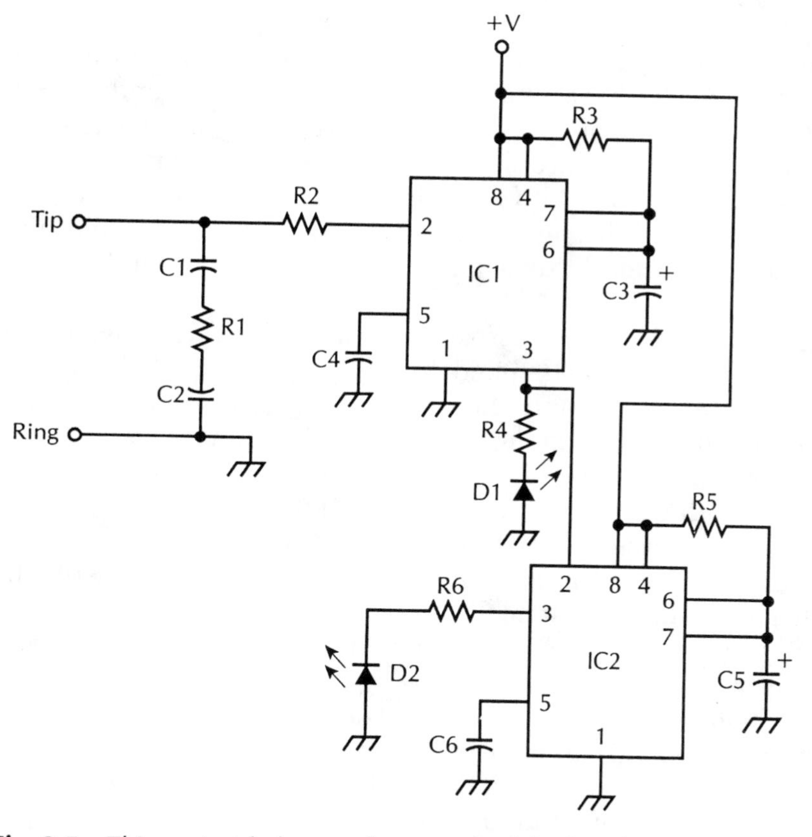

Fig. 9-5 *This project helps you keep track of the length of your telephone calls.*

Table 9-5 Parts list for the call timer project of Fig. 9-5.

IC1, IC2	7555 or 555 timer
D1	Green LED
D2	Red LED
C1, C2	0.22-μF capacitor
C3	See text
C4, C6	0.01-μF capacitor
C5	See text
R1	1-MΩ, 1/2-W, 5% resistor
R2	220-kΩ, 1/2-W, 5% resistor
R3	See text
R4, R6	330-Ω, 1/4-W, 5% resistor
R5	See text

multivibrator stages, each built around a common 555 or 7555 timer chip. The 7555 is a CMOS version of the standard 555. These two devices are pin-for-pin compatible. You can use either one without making any changes in the external circuitry. Generally speaking the 7555 gives slightly better performance than the 555, in terms of timing accuracy and low power consumption, but the 555 certainly does an acceptable job. The choice is up to you.

You might want to use a 556 dual timer IC in place of the two 555s (or 7555s) assumed in Fig. 9-5. This will reduce the overall size and possibly the cost of the project slightly. Such a modification is easy enough to make—all you have to do is correct the pin numbers appropriately. For your convenience, this version of the circuit is shown in Fig. 9-6.

Notice that an external, dedicated power supply is used for the main body of this circuit. Do not attempt to run it off the voltage on the telephone lines or you might cause excessive loading of the lines, disrupting normal telephone service and possibly subjecting you to severe legal penalties.

The power supply requirements for this project are not critical. The 555 (and its relatives) functions on any supply voltage from 5 V up to about 15 V. For best results, I recommend using a supply voltage of either 9 V or 12 V.

Like most of the projects in this book, this circuit is connected across the ring and tip wires of the telephone lines, as if it was an extension telephone. Any ac component in the signal

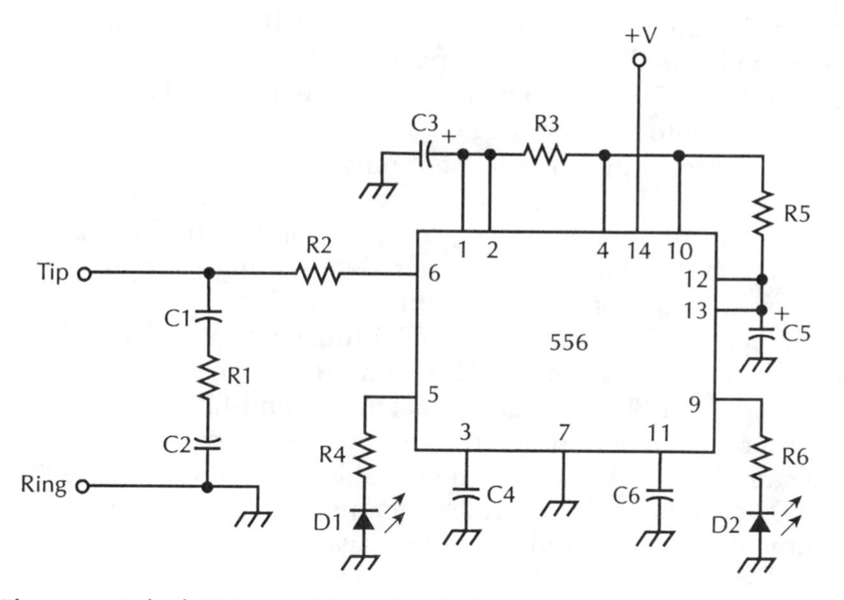

Fig. 9-6 *A dual 556 timer IC can be used in place of the two separate 555 timers in the circuit of Fig. 9-5.*

(including the ring signal is shunted across capacitors C1 and C2 and resistor R1. An ac signal will not get through to the circuit. Only a dc voltage can pass through resistor R2 to pin 2 of IC1. This is the trigger input of the first monostable multivibrator stage.

The 555 is triggered by a high-to-low transition. Ordinarily, the telephone lines carry about +48-V dc when the telephone is on the hook. This is seen as a high-level signal by the timer, even with the voltage drop across resistor R2. The timer ignores this continuous high signal, and its output remains low. Both LEDs in the circuit are dark and nothing happens.

Lifting the handset (whether to answer the ringing telephone or to place an outgoing call) drops this voltage to its off-hook level, a little less than +10 V. This looks like a low to pin 2 of IC2, triggering the timer. You might have to experiment with the value of resistor R2 to get reliable triggering in this project.

Once the timer is triggered, its output goes high, lighting up LED D1. Resistor R4 is a current-limiting resistor included to prevent the LED from drawing excessive current and damaging itself. The output remains high for a period determined by the values of resistor R3 and capacitor C3, according to the simple, standard formula

$$T = 1.1RC$$

where T is the time in seconds, R is the value of resistor R3 in ohms, and C is the value of capacitor C3 in farads. Do not confuse the units. For the moment I will ignore the exact time period and move on to the next stage.

The output (pin 3) of IC1 not only drives LED D1, it also is fed into pin 2 of IC2. This is the trigger input of the second monostable multivibrator stage, which is identical to the first (except for the values of the time-determining components—resistor R5 and capacitor C5 in this case).

When the first timer stage (IC1) times out, its output drops from high to low, triggering the second timer stage (IC2). LED D1 goes dark, and LED D2 lights up. This second LED continues to glow for a time period determined by the values of resistor R5 and capacitor C5. When this monostable multivibrator times out, LED D2 switches off. Both LEDs remain dark until the telephone is hung up and the handset is lifted again.

In this application I recommend making LED D1 green and LED D2 red. While talking on the telephone, as long as the green LED is glowing, you know you're okay. When the green LED goes out and the red LED lights up, you know it is time to wrap up the conversation. When both LEDs go dark, you know you have talked too long.

Notice that you don't have to watch the LEDs continuously or consciously keep track of the time. Just glance at the project periodically during the conversation. By noticing which (if either) LED is lit, you know the status of your call as far as time goes. There are just three possible states: green light—okay to keep talking; red light—time to stop talking; and no light—you've talked too long.

If both the red and green LEDs are simultaneously lit, there is something seriously wrong with the circuitry in your project. This should never happen.

Now all you have to do to complete the project is to determine the appropriate values for the time-determining components (R3, C3, R5, and C5) in the circuit. Obviously these component values depend on your decision about the maximum length of a telephone call. The first timer determines the length of the main body of the call. The second timer determines the warning (red light) time period. For most applications, a 1-minute warning period is about right. Many combinations of component values will give you a timing period of about 1 minute. A good combination is a 100-μF capacitor for C5 and a 560-kΩ resistor for R5. Ignoring component tolerances, this gives a timing period of slightly more than 1 minute.

$$\begin{aligned} T &= 1.1\text{R5C5} \\ &= 1.1 \times 560{,}000 \times 0.0001 \\ &= 61.6 \text{ seconds} \end{aligned}$$

In this application, will an extra second and a half really matter? I doubt it.

This means the total length of the call is equal to IC1's timing period plus 1 minute. Let's say you want to limit your telephone calls to 5 minutes. Therefore, you want the first monostable multivibrator stage to have a timing period of about 4 minutes (240 seconds). Added to the 1-minute warning period (IC2), this gives us a total of 5 minutes. Remember, the timing equation is

$$T = 1.1\text{R3C3}$$

You can choose a likely value for capacitor C3 and rearrange the equation to solve for R3:

$$\text{R3} = T/1.1\text{C3}$$

Let's try using a 470-μF capacitor for C3. In this case the ideal required value for resistor R3 would be

$$\begin{aligned} \text{R3} &= 240/(1.1 \times 0.000\,47) \\ &= 240/0.000\,517 \\ &= 464{,}217\ \Omega \end{aligned}$$

You can round this off to the nearest standard resistor value, or 470 kΩ (470,000 Ω). Returning to the original equation, you get an actual timing period (ignoring component tolerances, of course) of

$$\begin{aligned} T &= 1.1 \times 470{,}000 \times 0.000\,47 \\ &= 243 \text{ seconds} \end{aligned}$$

Again this should certainly be close enough. Is anyone going to be upset because the project permits calls to last 4 minutes and 3 seconds, instead of 5 minutes exactly?

Let's try another example. This time, let's permit a 10-minute call, with a 1.5-minute (90 seconds) warning period. The first monostable multivibrator stage should be set up for a timing period of 8.5 minutes (510 seconds).

Use a 1000-μF capacitor for C3 and a 220-μF capacitor for C5. First, solving for resistor R3:

$$\begin{aligned} R3 &= 510/(1.1 \times 0.001) \\ &= 510/0.0011 \\ &= 463{,}636\ \Omega \\ &= 470\ \text{k}\Omega \end{aligned}$$

Then solving for resistor R5, you get a value of

$$\begin{aligned} R5 &= 90/(1.1 \times 0.000\ 22) \\ &= 90/0.000\ 242 \\ &= 371{,}901\ \text{k}\Omega \end{aligned}$$

You can use either a 390-kΩ resistor or even a 330-kΩ resistor for R5.

Capacitors C4 and C6 are included to stabilize the timer ICs. The exact value of these capacitors is not critical. They might not be needed in all cases, but I consider them cheap insurance against an erratically performing project.

If you are using this project to control a rebellious teenager or someone else who can't be trusted to cooperate, you might want to modify the project. You can design a circuit to automatically hang up (disconnect the call) when IC2 times out. But this can be a problem if an emergency situation requiring a longer telephone call occurs.

Another solution is to sound an audible alarm when IC2 times out. This alarm can be a repeating beep that is applied to the telephone lines so both parties on the call can hear it. This should not be a very loud signal, but it should be rather annoying to serve its purpose.

❖10 Testing telephones

If you work with telephone systems, it is inevitable that sooner or later you are likely to run into problems. You need some ways to test a questionable telephone.

As noted in the first chapter of this book, under deregulation, it is now the individual consumer's responsibility to maintain and repair any telephone-related equipment they own. Before deregulation, virtually everyone rented their telephone equipment from Ma Bell, so AT&T handled all maintenance responsibilities, and customers enjoyed free service on both the telephone lines and their telephone equipment. For example, if you dropped your telephone and it broke, the odds were that Ma Bell would repair or replace it, often with no charge to you.

Those days are long gone. If you own your own telephone equipment, you certainly can't expect Ma Bell to repair it for free. It's your equipment, not theirs. So it's your responsibility, not theirs. Most telephone companies will not even send a repairman out to deal with a broken telephone. They are only responsible for their own lines. In most cases, the telephone company's responsibility for maintenance ends where the wires enter your house (or office building, or store, or whatever). From then on, you are on your own. They will send service personnel to troubleshoot problems with the wiring inside your building, but you can certainly count on paying quite a bit for this extra service. Except in highly unusual circumstances, where you have directly caused some sort of physical damage, a defect in the wiring outside is not considered your responsibility, and you will not be charged for such repairs.

Most local telephone companies now offer their customers, for a few extra dollars per month, some form of maintenance insurance to cover indoor wiring. Because it is usually a fairly simple matter to maintain the indoor lines on your own, this is a questionable bargain. Most consumers never have a problem with their indoor lines, and most problems are rarely difficult for anyone familiar with electronics to troubleshoot and repair themselves. If you are capable of building any of the projects in this book, you should be able to deal with 95 percent of the maintenance problems you are ever likely to encounter with your indoor telephone wiring.

This chapter gives you some tips on maintaining, troubleshooting, and repairing your telephone lines and equipment. Generally speaking, this is not difficult because telephone systems are usually relatively simple, especially when compared to other types of electronic equipment. Servicing a telephone is a far less daunting task than servicing a television set or a computer.

Simple logical troubleshooting

In most cases you don't need any special equipment—just a little common sense and logic. For example, let's say a telephone is completely dead. Is the problem in the telephone itself or somewhere in the telephone lines? If you have a second telephone, which you know to be good, unplug the original telephone and plug in the substitute. Does it work? If so, then the problem is in the original telephone unit. It must be repaired or replaced.

On the other hand, if the second telephone doesn't work, the odds are good that the problem is in the telephone lines. The telephone is not receiving the necessary incoming signals. It is unlikely that both telephones would go dead at once. However, strange things do happen. To double-check, try one or both of the telephones on another telephone line, preferably one that is known to be good. For example, take your questionable telephone to a friend's house. If his telephone is working, you know his line is good. Plug your telephone into this line. If it works, you know the telephone unit itself is okay, and the problem is in your telephone line. On the other hand, if your telephone still doesn't work on your friend's line, then you know you have a defective telephone unit. Notice that even in this case, you might also have a problem with the incoming telephone lines in your home.

This rather crude, hit-or-miss type of testing is sufficient in some cases, but it is hardly a very sophisticated approach. It doesn't tell you too much about the actual problem—just whether it is in the telephone unit or the telephone lines.

For more sophisticated and detailed testing, you need some sort of test equipment. In most cases, a standard volt-ohmmeter (VOM) or digital multimeter (DMM) can handle most of the measurements. As with any other area of electronics servicing, specialized test equipment can make the job a lot easier and more efficient. This chapter presents several simple projects designed to aid in testing telephones and indoor telephone lines.

I should mention that if you have an outdoor extension telephone or remote ringer, it still counts as part of the indoor wiring. When I say indoor wiring, I am referring to everything after the point where the telephone company's lines enter your house, office, or store; that is, everything on your side of the main connection block.

Telephone test signal source

For accurate and meaningful tests on telephones, you need a source of controlled test signals. The circuit shown in Fig. 10-1

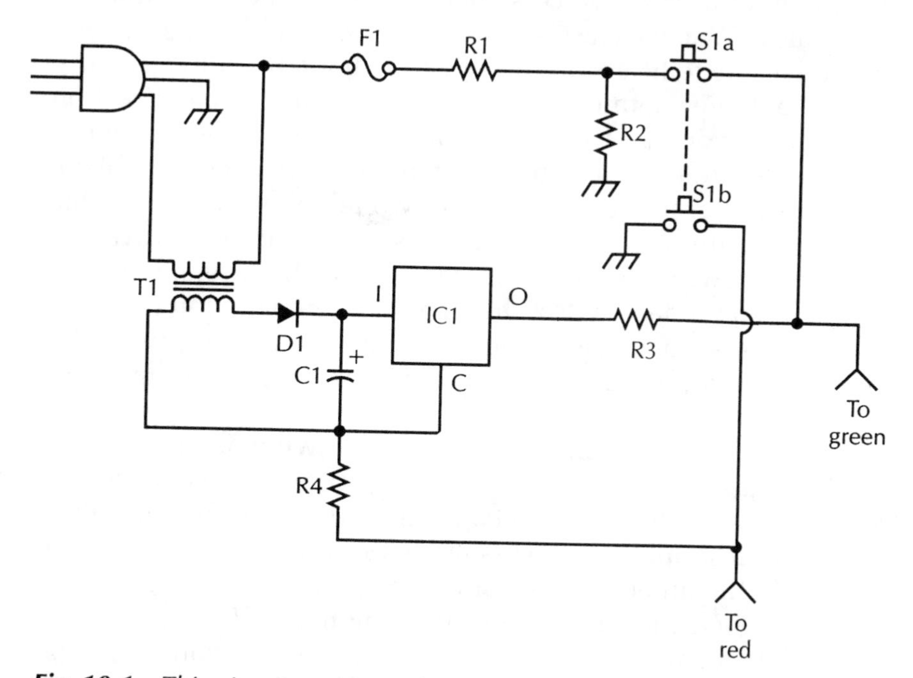

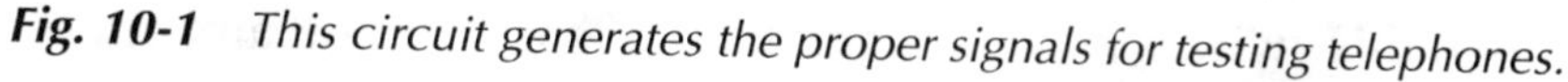

Fig. 10-1 *This circuit generates the proper signals for testing telephones.*

Table 10-1 Parts list for the telephone test signal source project of Fig. 10-1.

IC1	LM7812 12-V regulator
D1	Diode (1N4004 or similar)
T1	18-V ac transformer
F1	1/4-A fast-blow fuse and holder
C1	1000-μF, 100-V electrolytic capacitor
R1	3.3-kΩ, 1/2-W, 5% resistor
R2	8.2-kΩ, 1/2-W, 5% resistor
R3, R4	470-Ω, 5-W, 5% resistor
S1	DPST normally open push-button switch (A DPDT switch can be substituted; see text.)
	ac plug

is very useful in testing telephones. A suitable parts list for this project is given as Table 10-1. The component values in this project aren't too critical, although it is advisable to keep them reasonably close to the values suggested.

To use this telephone tester circuit, you need to hook two telephones to it in parallel. The circuitry provides the necessary operating voltages for voice signals to be sent between the two telephones. When an assistant speaks into telephone A, you should be able to hear him through telephone B, and vice versa. If the telephone communication does not work in one direction or the other, you can trace the supply voltages through the circuitry of each telephone to track down the source of the problem. Because conversations between the two telephones are possible with this circuit, it can also be used as a wired intercom system.

Closing switch S1 for a moment causes both telephones to ring. If you only need to test the ring function, you don't need to hook up the second telephone. In some cases, using two telephones in parallel causes a lower than normal ring volume, but this is rarely a problem.

Notice that a momentary-action DPST switch is required for S1. You can substitute a DPDT switch, just make no connection to the normally closed contacts. Do not attempt to use an SPST switch in this application. This tends to throw off the operating voltages in the other portions of the telephone tester circuit.

It is a good idea not to hold the ring button (S1) closed for too long. Remember, you've got a rather large ac voltage going through this switch. If there is any kind of short circuit anywhere in the tester or in one of the connected telephones, this ring voltage can do quite a bit of damage. For this reason, the 1/4-A fuse (F1) is an absolute must in this circuit. Do NOT omit this fuse. The potential risk is too great not to spend 50 cents or so for a fuse. It is cheap insurance. Do not use a slow-blow fuse. It must be the fast-blow type to offer adequate protection.

Because this project uses ac power from your household power lines, you must use all normal precautions throughout. Be very careful to avoid any possible short circuits. Check and double-check your wiring before plugging your project into the wall socket. Make sure all connections are fully insulated before applying power to the circuit. Please don't risk a dangerous and potentially fatal electrical shock or fire hazard. The circuit must be enclosed in an insulating housing of some sort.

Before connecting the telephones, it is a good idea to check your telephone tester circuit with a voltmeter. You should read about 12-V dc between the green (positive—ring) and red (common—tip) connections. This is with switch S1 in its normal open position. When this switch is closed, you should get about 85-V ac (assuming the line voltage is 120-V ac, although there might be some fluctuation). If the ring voltage is too low, try reducing the value of resistor R1.

For greater convenience, wire two standard modular telephone jacks in parallel to the green and red wire output lines of this tester circuit. Do not worry about the yellow or black wire connections to the modular jacks, they are unused here.

Telephone line tester

The telephone company is responsible for their own cables and equipment, but their maintenance responsibilities end at the termination block where their cables enter your individual home (or office, or whatever). You must bear the expense of any trouble that occurs on your side of the termination block. The telephone company will send technical personnel out to troubleshoot and repair your indoor telephone lines, but you must pay for this service and it is usually quite expensive. If you're smart, you'll make very sure you need their assistance before you call them.

This project will help you do your own troubleshooting. The circuit is shown as Fig. 10-2. A suitable parts list for this project is given as Table 10-2.

This circuit plugs into a standard modular jack and gives you an indication of the various important operating signals ap-

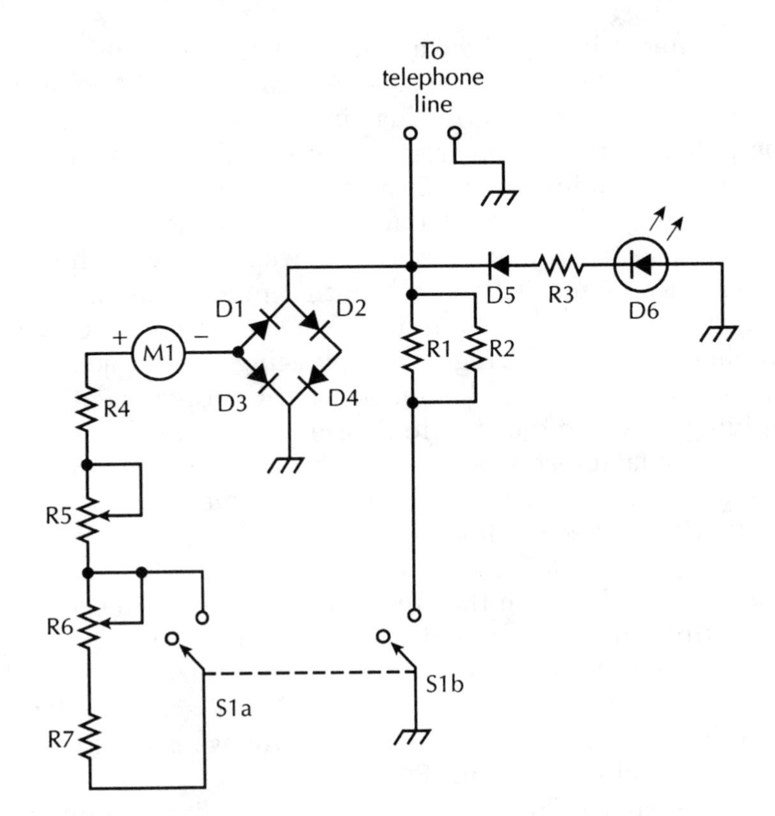

Fig. 10-2 *This circuit can be used to check the operating signals on your local telephone lines.*

Table 10-2 Parts list for the telephone line tester project of Fig. 10-2.

M1	1-mA dc milliammeter
D1–D5	Diode (1N4004 or similar)
D6	LED
R1, R2	470-Ω, 1/2-W, 5% resistor
R3	10-kΩ, 1/2-W, 5% resistor
R4	4.7-kΩ, 1/2-W, 5% resistor
R5	10-kΩ trimpot
R6	50-kΩ trimpot
R7	100-kΩ, 1/2-W, 5% resistor
S1	DPST (or DPDT) switch (see text)

pearing on the telephone lines. The test results are indicated on a 1-mA ammeter. For rough testing, if the pointer moves to the upper half of the meter's scale, the tested signal is assumed to be good. A current reading below 0.5 mA usually indicates some sort of problem.

If the circuit is hooked up backwards, with the wrong polarity across the telephone lines, the LED (D6) will not be illuminated. In normal testing procedures this LED should light up when the circuit is connected to the telephone lines. It might glow steadily or blink, depending on the signal being tested. If the LED does not light up when you plug the tester in, unplug it, and reverse the wiring to the plug. The project should work properly now.

You might find your telephone jack is wired backwards. Usually this doesn't matter, but a few electronic telephones are polarity sensitive and will not work when connected to a reverse-wired modular jack.

This tester is a fully passive device; it has no power source of its own. It "steals" its operating voltage from the telephone lines. Its current consumption is very low, so it should not affect the normal operation of the telephone lines.

Switch S1 is a DPST switch. It has two SPST sections, which always operate in unison. True DPST switches are rather difficult to locate, but a DPDT switch will work just as well; you'll just be left with two unused contacts.

When the switch is open and the tester is connected to a live modular jack, it will test the line voltage. All extension telephones (and other related devices) on the line must be in the on-hook condition. The on-hook voltage on the telephone lines should be at least 40-V. Because of Ohm's law, the unknown voltage can be measured through the fixed circuit resistance on a milliammeter. For a 1-mA meter, any reading above 0.4 mA indicates a good line voltage. The LED should glow continuously during this test.

What happens if the telephone rings during the test? You'll be helped out, because you can perform the ringer test. Or you'll have to arrange for a friend to dial your telephone number at a prearranged time. Most local telephone companies have special ring-back codes that permit technicians to dial a number to ring the telephone they dialed from for testing purposes. Contact your local telephone company's maintenance department to find out how to dial the ring-back number for testing your telephone. Be sure to speak to someone in the technical or maintenance department. I've found that operators and billing office personal either don't know what the ring-back code is or don't know what you're talking about.

To test the ringer signal, switch S1 must be open, as in the line voltage test. The milliammeter should read at least 0.5 mA (and probably significantly higher) and the LED should flicker. It is lit only when the telephone is actually ringing. During the between-ring pauses, the LED does not receive an ac signal, so there is no flickering during these pauses.

The exact reading you get on the meter during the ringer test is affected by the number of ringers (or equivalents) connected to the line. The more extension telephones and/or projects you have hooked to your telephone line, the lower the ringer voltage indicated by the milliammeter.

To test the off-hook line voltage, you do not pick up the handset of your telephone. You won't get any meaningful reading that way. Leave all telephones on the line on the hook, and close switch S1 on the tester. The circuit now simulates an off-hook telephone and gives an appropriate reading on the milliammeter. Once again, a good reading is 0.4 mA or higher. The LED might not light up during this test. This is normal and acceptable.

If you lift the handset of a telephone hooked to the line, the meter's reading should drop noticeably. If it does not, the telephone in question is probably defective.

Trimpots R5 and R6 can be used to calibrate the unit. Set both trimpots to about the middle of their range and hook the circuit up to a known dc voltage source of about 40 V. Adjust the trimpots alternately until the meter reads 0.3 to 0.4 mA. These two controls interact, and in some cases, the calibration procedure can be a little fussy and time-consuming. For simple yes/no testing (which is all the ordinary layman is likely to need), you don't have to worry too much about exact calibration. As long as the readings are reasonably close, the device will do the job just fine.

For more precise calibration, close switch S1 and feed in a well-regulated 5-V test signal. Adjust trimpot R5 for a reading of 0.3 mA. Then open the switch and increase the input voltage to 40 V, adjusting trimpot R3 until the meter's pointer is at 0.3 mA. Again, these two trimpots are likely to interact, so you might have to switch back and forth between them to get a truly precise calibration.

Don't expect miracles. The test equipment presented in this chapter is crude and nowhere near professional standards. But it should be quite sufficient to help you pinpoint the source of your problem and tell you if you need to call in a telephone company technician or send your telephone equipment in for factory authorized servicing. You won't end up paying for a service call only to be told, "Nothing wrong here, the problem must be something else in the system."

Index